Computer Vision
for
Electronics Manufacturing

ADVANCES IN COMPUTER VISION AND MACHINE INTELLIGENCE

Series Editor: Martin D. Levine
McGill University
Montreal, Quebec, Canada

COMPUTER VISION FOR ELECTRONICS MANUFACTURING
L. F. Pau

SIGMA: A Knowledge-Based Aerial Image Understanding System
Takashi Matsuyama and Vincent Shang-Shouq Hwang

Computer Vision
for
Electronics Manufacturing

L. F. PAU
Technical University of Denmark
Lyngby, Denmark
and University of Tokyo
Tokyo, Japan

PLENUM PRESS • NEW YORK AND LONDON

Library of Congress Cataloging in Publication Data

Pau, L. F. (Lewes F.)
    Computer vision for electronics manufacturing / L. F. Pau.
        p.   cm. — (Advances in computer vision and machine intelligence)
    Includes bibliographical references.
    ISBN-13:978-1-4612-7841-2   e-ISBN-13:978-1-4613-0507-1
    DOI: 10.1007/978-1-4613-0507-1

    1. Electronic apparatus and appliances — Design and construction. 2. Computer vision —
Industrial application. I. Title. II. Series.
TK7836.P375   1989   89-22910
621.381 — dc20   CIP

Cover illustration: A microscopic view of a mask image
after thresholding and two-step erosion, as seen
by the KLA 2020 Wafer Inspector. *Photo
courtesy of KLA Instruments Corporation.*

© 1990 Plenum Press, New York
Softcover reprint of the hardcover 1st edition 1990

A Division of Plenum Publishing Corporation
233 Spring Street, New York, N.Y. 10013

To my daughter, Isabelle Pau

Acknowledgments

Proper acknowledgments, with thanks, are hereby given to the following publishers, authors, and companies for permission to reproduce parts of their publications or photographs:

Publishers	*Science* (Ref. 237) (Figure 36)
	IEEE Spectrum (Ref. 84) (Figure 90)
	ACM (Ref. 149) (Figures 144 and 145)
	Electronics Manufacture & Test (Ref. 81) (Table 29 and Figures 81–85)
	Test and Measurement World (Ref. 7) (Table 7)
	Semiconductor International (Refs. 40, 68 and 70) (Table 18 and Figures 66 and 74)
	John Wiley and Sons (Ref. 20) (Figure 20)
Authors	C. Pynn (Ref. 41) (Table 21)
Companies	AdeptVision Inc. (Figures 113, 131, and 136)
	C. E. Johansson AB (Figure 115)
	Cambridge Robotic Systems Inc. (Figures 86 and 87)
	KLA Instruments Corporation (Figures 27, 60, 62, 63, 64, 70, 71, and 72)
	Cambridge Instruments (Figures 58, 65, 69 and 73)
	Cognex Corporation (Figures 93 and 96)
	Applied Intelligent Systems (Figures 31 and 61)
	Leitz (Figures 29 and 41)

Riber S.A. (Figure 35)
Carl Zeiss (Figures 15 and 16)
Morgan Semiconductor (Figure 53)
Sonoscan (Figures 102 and 103)
International Robomation Intelligence (Figures 109 and 112)
Photo Research Vision Systems (Table 5)
Texas Instruments (Figure 79)
Nicolet Instrument Corporation (Figure 91)
IVS/Analog Devices Corporation (Figures 94, 95, and 97)
Vanzetti Corporation (Figures 107 and 108)

This book is the result of approximately 15 years of work in the area, and mostly of collaboration with industry in the United States, West Germany, France, and Japan. It was written in 1987–1988 while the author was holding the CSK Chair in Information Science at the University of Tokyo (he was the first non-Japanese to do so).

All due thanks are hereby extended to Lene Duvier, Copenhagen, for the typing and preparation of the manuscript, in spite of transcontinental distances and tight schedules.

L. F. Pau

Contents

II. Vision Algorithms for Electronics Manufacturing

13. Image Quantization and Thresholding 207

14. Geometrical Corrections 215

15. Image Registration and Subtraction 221

Introduction and
Organization of the Book

COMPUTER VISION FOR ELECTRONICS MANUFACTURING

Regardless of the specific application or context, why is the role of computer vision so essential in electronics manufacturing? The following prime reasons must be put forward:

1. The ever-decreasing physical dimensions of all parts, patterns, and defects; the tight tolerances; and the higher layer counts make it increasingly impossible for humans to carry out inspection and parts measurement, at least with the speed required by manufacturing throughput (Figure 1).
2. Computer vision is mandatory as an in-process "auditing" facility, to help understand why, how, and where the defects or quality losses occurred, and ultimately in order to improve upon the process itself.
3. The trend toward application of specific designs, with smaller production runs of each item, gives versatile vision systems a major advantage over single-purpose test systems.
4. The processing power and price of specific or generic vision systems for electronics manufacturing have finally fallen into more reasonable ranges, especially as increasing portions of the inspection algorithms can be executed economically by fast vision-dedicated hardware (digital signal processors, array processing chips, pipelined processors, concurrent/parallel processors, and systolic and other arrays).

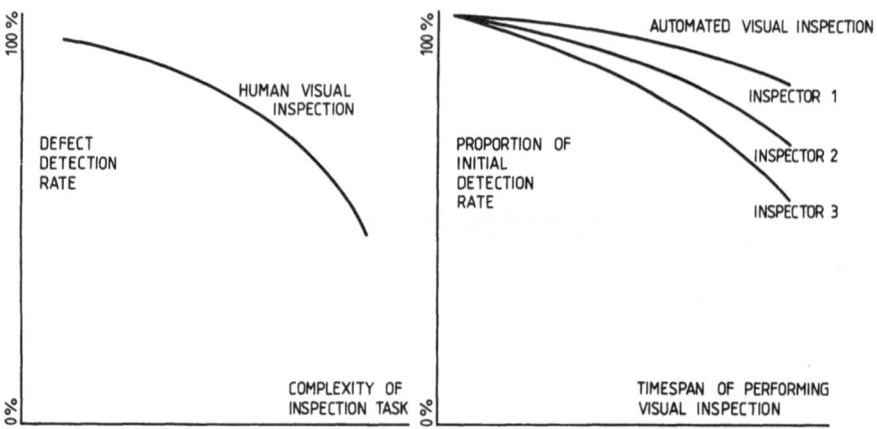

Figure 1. Trends in relations between the complexity of inspection tasks, defect detection rates (absolute and relative), and inspection time.

Irrespective of the necessities described above, and with the exception of specific generic application systems (e.g., bare-board PCB inspection, wafer inspection, solder joint inspection, linewidth measurement), vision systems are still not found frequently in today's electronics factories. Besides cost, some major reasons for this absence are:

1. The detection robustness or accuracy is still insufficient.
2. The total inspection time is often too high, although this can frequently be attributed to mechanical handling or sensing.
3. There are persistent gaps among process engineers, CAD engineers, manufacturing engineers, test specialists, and computer vision specialists, as problems dominate the day-to-day interactions and prevent the establishment of trust.
4. Computer vision specialists sometimes still believe that their contributions are universal, so that adaptation to each real problem becomes tedious, or stumbles over the insufficient availabllity of multidisciplinary expertise. Whether we like it or not, we must still use appropriate sensors, lighting, and combinations of algorithms for each class of applications; likewise, we cannot design mechanical handling, illumination, and sensing in isolation from each other.
5. Vision systems must be able to handle quickly and without rejection both the good parts and the bad ones, which are the ones usually highlighted in systems architectures.

The purpose of this book is to serve as a reference when specialists

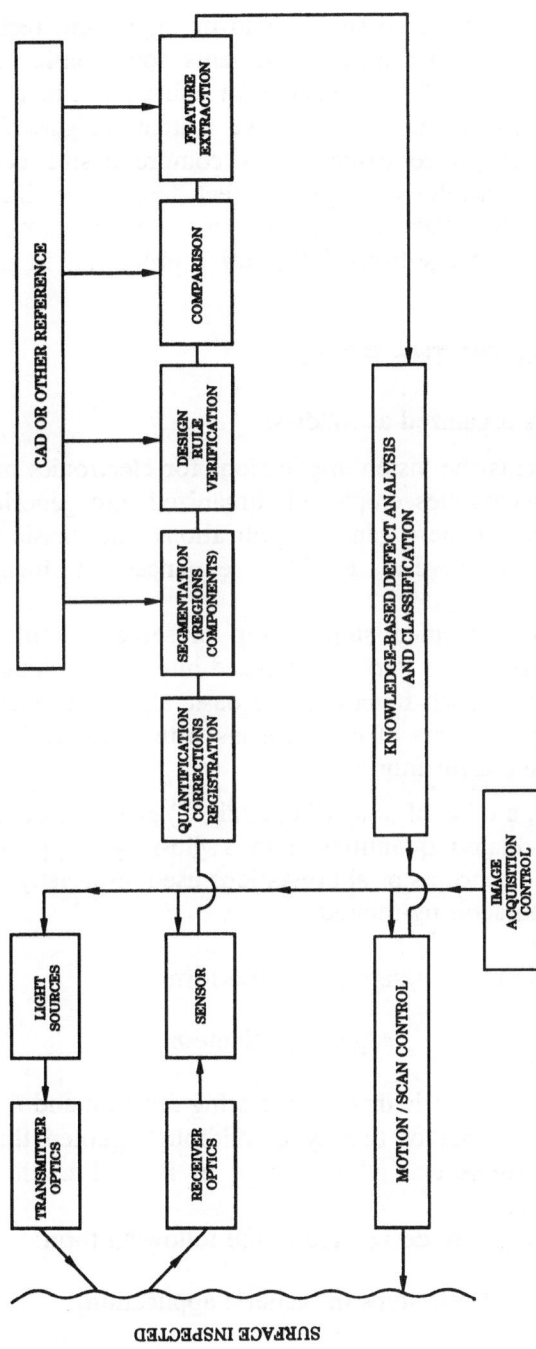

Figure 2. Typical functions in inspection tasks for electronic manufacturing.

from one discipline need to specify and/or implement techniques from the others. At the same time, it presents some basic and advanced techniques at a level of detail sufficient for a first implementation, and for considering trade-offs among alternative systems (Figure 2). The book does *not,* however, give extensive or comprehensive descriptions of physical sensor principles or design, defect analysis or characterization, image processing hardware architectures, more detailed image processing algorithms, or knowledge-based information processing in general.

ORGANIZATION OF THE BOOK

The book is organized as follows:

Part I presents the vision applications for electronics manufacturing and the basic techniques employed, organized into generic application classes. For most of these generic applications, the "basic steps" tables give one out of several feasible sequences of image-processing algorithms.

Part II gives detailed step-by-step presentations of each of the algorithms referred to in Part I, organized into generic image processing functions. For each such function, one basic algorithm is given, without any exhaustivity in terms of comparisons with other variants or refinements to the same algorithm.

In addition, a table of units (Appendix C) gives the definitions of the major imaging-related quantities, and a glossary (Appendix A) gives definitions of the concepts or abbreviations used, especially those that are process- or manufacturing-related.

The algorithms are designated in the form

Algorithm Name-n

where Name is the generic image processing function and n a specific (or not) algorithm number of that type. Although quoted throughout the book, the algorithms can all be found in Part II in the appropriate sections.

The basic steps are designated in the following form:

Basic steps in [generic application]

and incorporated as tables into the relevant chapters. The algorithms mentioned in the basic steps may be specific (e.g., Algorithm Edge-3) or generic when a choice is possible (e.g., Algorithm Edge-n).

Part I

Applications and Systems Aspects

Chapter 1

Vision System Components

This introductory chapter summarizes basic information on sensors, lighting, vision system specification, and general issues which is useful throughout the book; none of the sections is specifically related to applications of vision systems in electronics manufacturing. Therefore, readers interested only in these applications may jump to Chapter 2.

1.1. VIDEO SENSORS

There are several types of cameras that can be used as sensors in vision systems. Most often, video or charge transfer device (CTD) sensors are used (Table 1), but there are also other sensors that find use in special applications (Table 2); these include image intensifiers and image dissectors.

1.1.1. Video Cameras

In video cameras, electrons traveling from a heated element through vacuum to a light-sensitive faceplate are deflected on the way to form a raster pattern. The deflection is controlled with two oscillators, called line and field-timebase generators. At any given moment, the electron beam "illuminates" only a small spot on the faceplate. This causes a small area of the faceplate to be discharged and generates a signal proportional to the light falling on the target. This signal is amplified and mixed with

Table 1. Typical Performances of CCD and Vidicon Sensors

Parameter	CCD		Vidicon camera	
	Typical value	Max. value	Typical value	Max. value
Resolution (TV lines)	250–350	1000	600	2000
Spatial bandwidth (MHz)	1	6	10	30
Sensitivity (lum.) $(V/\mu W \cdot cm^{-2})$	0.08	Variable	0.01	0.02
Dynamic range (peak signal/RMS noise)	2000:1	10,000:1	1000:1	1200:1
SNR (db)	54–59	76	60–63	70
Geometric distortion	0	10^{-3} pixels	1–2% of total frame	2%
Readout speed (frames/s)	25(30)	400	25(30)	2000
Nonuniformities (%)	2	12	10% between center and corner	20%
Lag (residual current after 200 ms) (%)	0	Variable	10	15–20
Spectral sensitivity (nm)	300–1200	8000	200–14,000	80,000 (IR)

synchronizing signals to generate a composite video signal (see also Section 1.2).

1.1.2. Charge Transfer Devices (CTD)

Charge transfer device sensors are made of light-sensitive silicon. CTD sensors resemble metal-oxide-semiconductor field effect transistors (MOSFETs) in that they contain a "source" region and a "drain" region coupled by a depletion region channel. For imaging purposes, they can be considered as a monolithic array of closely spaced MOS capacitors.

CTD sensors can be divided into three groups on the basis of the technique used to read from the array: charge-coupled devices (CCD) with line transfer, CCD with frame transfer, and charge injection devices (CID). The first two types are also called line scan and area scan CCD cameras.

1.1.3. Shape of the Sensor

The shape of the sensor can be very important, especially when the object is moving. Video cameras have only one shape: array sensors. In CTD cameras, it is possible to select between different sensor shapes. The three most common types are array, line, and disk-shaped sensors. When the object is stationary, or slowly moving, an array can be used. For rapidly moving objects, improved performance is obtained by using line or disk-shaped sensors coupled to a motion synchronizer.

Table 2. Image Sensors and Sensing Principles

- Vidicon, TV (visible, IR) (coherent, incoherent, polarized, scattered)[a]
- CCD arrays (visible, IR) (coherent, incoherent) (linear and others)
- UV, IR, γ-ray sensors (coherent, incoherent)
- Mechanical gauging
- Pressure sensors or arrays (touch, tactile, sonar)
- Laser beams or arrays (single beams, multiple beams, scanning, diffraction)
- Range finders and triangulation (lasers, ultrasound, optical, depth from stereo)[a]
- Fiber optic vision sensors, endoscopes, and other sensors[a]
- Multicolor illumination (optical incoherent, multiple-wavelength lasers, and radars)[a]
- X-ray radiography and microscopy
- X-ray diffraction and fluorescence
- Ultrasonic sensors (A, B, C scans)
- Radars
- Neutron radiography
- Piezoelectric and capacitive tablets
- Flash imaging, stroboscopic imaging (mechanical or electronic shutter to avoid ghosting)
- Microwave sensors, imaging and tomography
- Phase conjugation with partially coherent light
- Inverse optical or microwave scattering (tomographic and projective reconstruction of detail, laser surface scattering)[a]
- Wavefields reconstructed from scaled holograms
- Interferometric imaging (optical, laser)
- Moiré technique
- Eddy-current signals or holography
- Acoustic emission
- Radiometric scanning
- Speckle methods (laser interferometry)
- Optical synthetic apertures[a]
- Triangulation and range image reconstructions from multiple images[a]
- Microwave and sonic synthetic aperture (ultrasonic, infrasonic, sonar, radar)
- Stereo with two or three sensors (visual, laser scan, and photometric)[a]
- Albedo reflectivity map[a]
- Extended Gaussian sphere
- Reflectance/reflectivity map technique, fringe projection (incoherent, laser)[a]
- Structured illumination (shadowing, grids, oblique Fizeau)[a]
- Magnetometer scanning
- Electron image processing (SEM, TEM)
- Ion-beam image processing
- Vibrothermography
- NMR imaging
- Luminescence

[a] Best for inference of 3-D information.

1.1.4. Sensor Resolution

Sensor resolution is the sensor's ability to distinguish detail in an image. Resolution is normally expressed as the number of television lines per picture, e.g., 625 lines over a 340-mm picture height. The most accurate way to measure resolution is with the modulation transfer function (MTF).

In CTD cameras, the limit is set by the number of pixels available in the image area.

In tube cameras, resolution is influenced by the type and size of the photoconductive layer, image size on the layer, beam and signal levels, optics, and spot size of the scanning beam.

1.1.5. Sensitivity

Sensitivity is the efficiency of the light-to-charge conversion in the sensor. There are several ways of measuring the sensitivity of a sensor; the two most frequently used are luminous sensitivity and radiant sensitivity:

1. *Luminous sensitivity* is measured at a specific color of light, usually 2856 K. The output is expressed in $mA/lm \cdot mm^{-2}$ or $V/mW \cdot mm^{-2}$.
2. *Radiant sensitivity* is measured over a range of wavelengths, usually from 400 to 1000 nm. The output is expressed in $mA/W \cdot mm^{-2}$.

Some producers give information about the sensitivity in terms of the minimum lumens of white light (expressed in lux) giving a readout higher than the noise (SNR = 1), and usually measured with a $f/1.4$ lens mounted on the camera.

It can be difficult to compare the sensitivity of cameras because of these different methods and units. The units are related through the equations

$$F(\text{lumens}) = c \int V(l)P(l)\, dl$$

$$\text{lux} = \text{lumens/mm}^2$$

where c is a constant, $V(l)$ is the sensor response in lumens/unit wavelength as a function of l, $P(l)$ is the radiant flux in watts/unit wavelength as a function of l, and l is the wavelength.

In CTD cameras, sensitivity is influenced by variables such as quantum efficiency, the length of integration time, and the dominant source of noise in the device.

Video camera sensitivity is dependent on the type and size of the photoconductive layer. It also varies with the target voltage level in certain layer types.

1.1.6. Dynamic Range

Dynamic range represents the overall usable range of the sensor. It is usually measured as the ratio between the output signal at saturation and the RMS value of the noise of the device (sometimes peak-to-peak noise). In CTD cameras, the RMS noise does not take into account dark-signal nonuniformities.

In CTD cameras, the saturation voltage is proportional to the pixel area.

Factors influencing dynamic range of video cameras include photoconductive characteristics of the layer as well as scanning rate and electronic gun characteristics.

1.1.7. Signal-to-Noise Ratio (SNR)

The signal-to-noise ratio measures similar things as dynamic range. Dynamic range is a measure of sensor characteristics, while SNR is a measure of the overall camera characteristics. SNR is measured in decibels (dB) (ratio of peak signal to RMS noise).

In video cameras, the main noise source is the preamp electronics, while in solid state cameras, the fixed pattern noise is the dominant source.

1.1.8. Geometric Distortion

The geometric distortion tells how much a sensor distorts geometrically an image (Figure 3). In CTD cameras, it is measured as a number of pixels (usually very small) with respect to the total number of pixels. In video cameras (where geometric distortion can be a problem), it is measured in percentage of lines with respect to full frame aperture. Correction algorithms are given in Chapter 14.

1.1.9. Readout Speed

Readout speed is the frequency of the frame transfer, measured in frame/s. Standard frame rate is 25 (30) frames/s, but it is possible to get slower (slow scan) and higher readout speeds. See also Section 1.2.

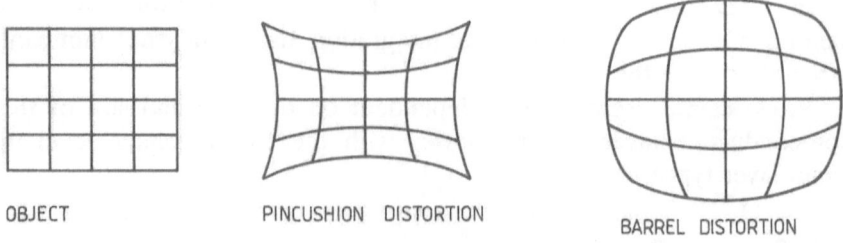

OBJECT   PINCUSHION DISTORTION   BARREL DISTORTION

Figure 3. Geometrical distortion effects.

1.1.10. Spectral Sensitivity

The spectral sensitivity gives the sensor's relative response at different light wavelengths (Figure 4). Producers usually give diagrams of this response. It is also possible to express this response by the position of the peak and a 6-db interval.

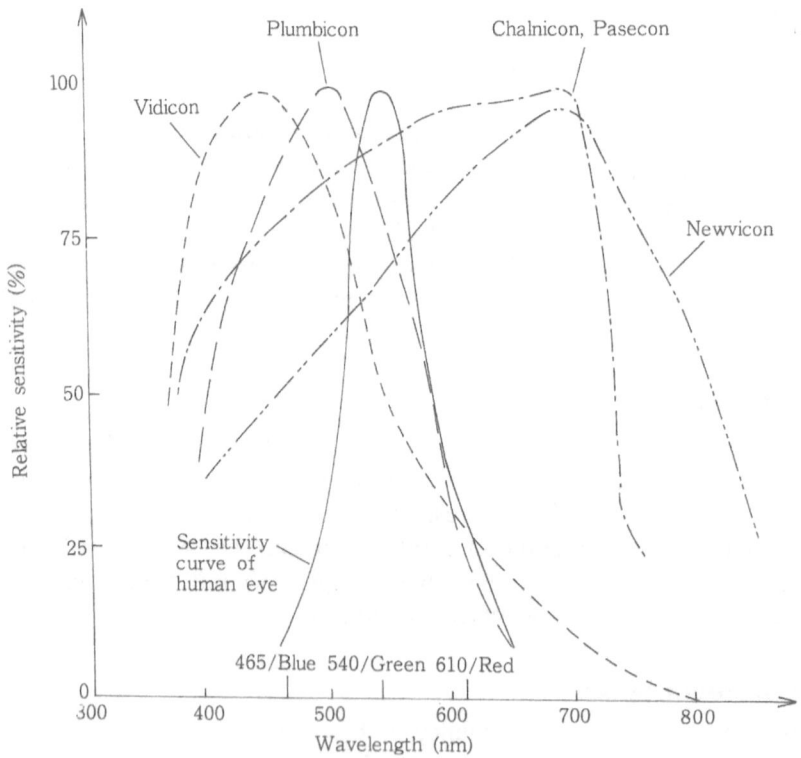

Figure 4. Spectral response of video sensors.

Usually, CTD cameras cover wavelengths in the range 0.3–$1.2\,\mu$m, whereas video cameras cover 0.2–$14\,\mu$m.

1.1.11. Lag

The lag is the time it takes for the signal output to change in response to a rapid change in the light level on the target. This rapid change can be an increase or decrease in brightness. Lag is measured in percentage of the residual charge with respect to the saturation charge.

CTD cameras usually have a very small lag, but lag can be a problem in video cameras. In video cameras, lag is influenced by the type of photoconductive layer used, its size, the image's signal level, and electron gun characteristics.

1.1.12. Camera Synchronization

There are several ways of controlling the synchronization (sync) of the camera (Figure 5). Gen-locking or sync-locking is the term used for locking a camera's sync generator to a computer. The sync generator supplies vertical and horizontal blanking, black level clamping, and other signals needed for camera operation. The sync generator can be triggered externally by a computer or another timing master. As a timing source for the sync generator, a phase lock loop, a direct clock (external) drive, or an internal clock can be used alternatively.

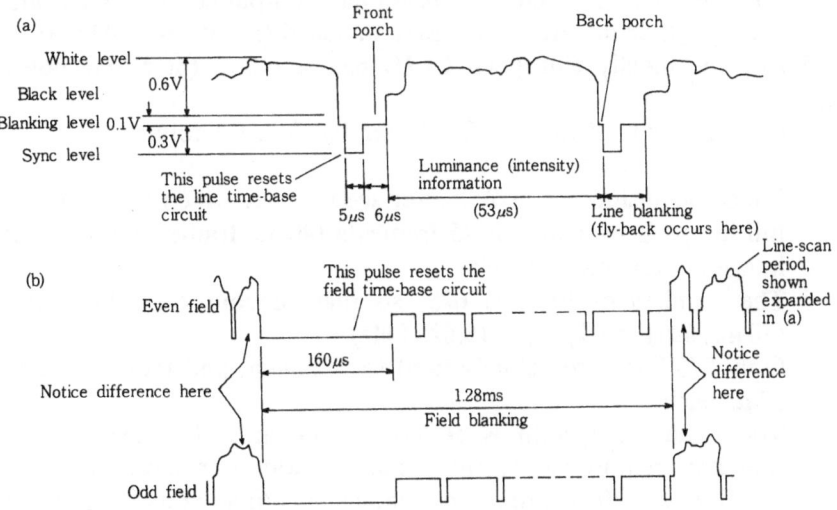

Figure 5. Structure of composite video signals (EIA standard).

The camera is driven by an internal mechanism, and it may also output sync signals to other cameras. Related strobing aspects are discussed in Section 1.4.

1.1.13. Nonuniformities

Video cameras tend to give higher signal output at the center of the frame than at the edges; they also suffer from random nonuniformities for thermodynamic reasons.

CTD cameras do not suffer from systematic nonuniformities; they do, however, have two kinds of random nonuniformities:

1. *Response nonuniformity* (*RNU*) exists both because of variations in the area of the element and from material and processing variations in the chip manufacturing. This effect is equivalent to a gain variation from element to element.
2. *Dark-signal nonuniformities* (*DSNU*) are due to materials variations across the chip and discrete defects in the sensor. This type of nonuniformity is very process dependent and varies greatly from chip to chip. It is not possible, therefore, to be confident of the level of DSNU from the manufacturer's data sheets.

1.2. CCIR 625 VIDEO STANDARD

Most video sensors applied to vision have output signals which obey unfortunately either the American EIA standard (NTSC at 60-Hz power supply) or the mostly European CCIR 625 standard (at 50-Hz power supply).

The main points of the CCIR 625 standard are the following:

- Image acquisition is in 2:1 interlaced mode, which means that a full image is available at 25 frames/s (40 ms/frame) and one-half image is available at 50 Hz.
- The number of lines is 625, so that readout of a line takes (40 ms/625) = 64 μs (or 15,625 kHz).
- Only 576 lines are actually used for imaging, and they are active 52 μs each.
- The signal bandwidth is at 5 MHz, meaning that according to sampling requirements, the signal readout per image point at 10 MHz lasts 100 ns; thus, the readout length in 52 μs is 520 pixels (or 12.8 mm for most video sensors).

- The stored image line length is only 512 pixels or 12.6 mm.
- The stored image line height is only 512 pixels or 8.53 mm instead of 576.

1.3. SCANNING OF THE DIGITIZED IMAGE

The video signal, once digitized (see Algorithm Quant-1), is stored in an image frame store, which itself can be retrieved in different ways as required by the image processing to take place from there on. The retrieval fetches the pixel values for all image points within a scanning grid.

Depending on the shape of the scanning grid, different spatial sampling properties are achieved. They are in terms related to the distances to the closest neighbors to each pixel in the grid:

Grid shape	Distances to neighbors (1 = pixel-to-pixel line distance; E = line-to-line distance)
Square	$1, \sqrt{2}, 2, \sqrt{5}$
Rectangular	$1, E, (E^2 + 1)^{1/2}$
Triangular	$1, \sqrt{3}, 2$
Hexagonal	$1, \sqrt{3}, 2$

It is fundamental, and often critical, to determine the scanning pattern, size, and resolution jointly as this determines the later signal-processing architectures and performances; also, the faster an object moves while being scanned by an array, the more smear or image degradation results from matrix scan time requirements. $N \times M$ arrays are useful for applications involving stationary or very slowly moving objects; for rapidly moving objects, improved performance is obtained by using $1 \times N$, radial, rosette, or other special arrays, coupled to a motion synchronizer. When combined, the two (matrix array combined with $1 \times N$, radial or rosette) constitute a frame whose size may be commensurate with the application requirements. The detector specifications also influence scanning-pattern requirements, e.g., in relation to matching and feature extraction.

In advanced designs, no complete scan is made, and the pattern is itself a control variable looking for expected features following a tree search procedure.

Sparse scanning may be necessitated by data-compression requirements, both for transmission and feature extraction. Coarse sampling is
also desirable to limit memory size, but some interpolation algorithm is
necessary, of which the best is linear interpolation with slight oversam-
pling (see Algorithm Geom-2).

If high spatial resolution is required, accentuation by a postfilter
often improves image quality (Algorithm LUT-1). A useful method here
is to compute the space-dependent error function due to the scanning,
sampling, and quantization, with interpolation/reconstruction by a low-
pass filter close to the ideal filter.

When some features are much smaller than the scan-to-scan distance
in the scanning geometry, the expansion–contraction method can be
employed (Algorithm Morph-2). Here, each pixel is enlarged in all
directions to obtain an enlarged pattern. This enlarged pattern is then
contracted by the same amount. Such processing eliminates small convex
features, and the rest of the pattern is approximately restored. The small
features will appear in an image subtraction between the raw picture and
the enlarged–contracted picture. Similarly, small concave features may
be detected by first contracting the input pattern. This scheme works
especially well for binary patterns.

1.4. STROBE LIGHTING

1.4.1. Introduction

The need for strobing arises mainly due to the fact that the objects to
be imaged are examined while moving. Strobing reduces the effects of
image blur that occurs while photons are accumulated by the vision
sensor during its finite scan period as the object moves through its field of
view. The solutions include either mechanical shutters or electronic
shutters; the latter either take the form of a light valve (e.g., liquid
crystal shutter) or expose the object to a short light pulse (typically,
under 1 ms from, e.g., a gas discharge tube) with the area sensor control
circuit controlling the short exposure time; in this case, the sensor is
called a monoshot camera.

1.4.2. Motion Freeze

According to the standard American EIA TV camera timing (see
Section 1.2), a television field is scanned in 16.7 ms; a television frame
consists of two interlaced TV fields and requires 33 ms, corresponding to
a 60-Hz frame rate. Therefore, the image blur is determined by the

velocity of the object being viewed according to the apparent motion D during its scan period:

$$D = V \times T$$

where V is the object velocity and T is the time taken by a complete scan (16.7 ms in a TV field, and much less for the strobe duration). The ratio of the apparent motion to the field of view, D/F, must be less than one pixel in the digitized image frame resolution (F is the field of view).

1.4.3. Flash Selection

The strobing light source can be characterized in terms of the following properties:

- Type (xenon flash, fluorescent tube, laser, flash X-ray).
- Energy in joules, and electrical-to-light-energy conversion ratio.
- Average power dissipated.
- Instantaneous power during the strobe duration.
- Pulse duration, related to the capacitance of the flash capacitor and resistance of the flash circuit.
- Emission spectrum; a xenon tube emits light over the range 0.2–2 μm, with peaks usually at 0.48 and 0.8 μm.
- Repetition rate, typically up to 30 Hz.

1.4.4. Strobe Synchronization

In order to obtain a complete image of the scene under strobe lighting, the strobe firing must be synchronized with the sensor and image acquisition, and sometimes also with reference CAD data retrieval. Typically, the strobe must be fired during the vertical blanking period between scan periods, where no video information exists (Figure 6). By firing the strobe during these approximately 2 ms, a complete image is stored by the sensor, then scanned and transferred during the next 16.7-ms field period.

Since the strobe can only be fixed at or during the vertical blanking period, the object position uncertainty D corresponds to one TV field, $T = 16.7$ ms. This causes problems if the field of view is small. Therefore, an alternative is to fire the strobe light immediately on object arrival, detected by beam photoeyes, or on end of transfer of reference CAD data; the TV sensor must then have scan inhibit and timing signals synchronization.

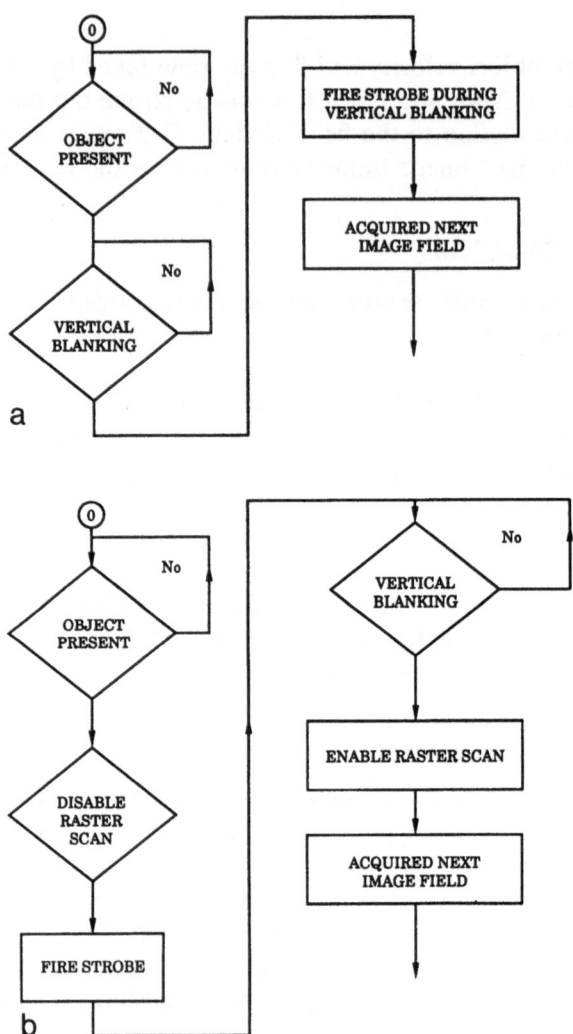

Figure 6. Strobe lighting synchronization: (a) vertical blanking strobe; (b) strobe and vertical blanking synchronization.

1.4.5. Strobe Beam Geometry Configuration

To exploit device geometries, surface reflectivities, and slopes, the strobe beam geometry must be optimized selectively to illuminate certain features.[1] This typically involves fiber optic channeling to specularly reflect off some surfaces or to illuminate specific points or lines. The latter especially applies to increase locally the contrast between an object, e.g., a component edge, and the background; such edges or lines are imaged when the strobe lighting is stepped through such lines one at a time while the sensor remains unchanged.

1.4.6. Reflection Elimination

The most extreme sensor signals occur for high-intensity strobed light on shiny surfaces where specular reflection occurs into and away from a collection lens. With unidirectional illumination, the lens must collect the reflected light over a wide range of angles. Conversely, if the illumination geometry is multidirectional, the collection lens's numerical aperture can be reduced, resulting in lower cost and greater depth of field.

When using polarization discrimination, the effect of surface characteristics must be thoroughly understood. In case of a wider polarization mix of directions in the reflected than in the incident lighting, circularly polarized light may give a more balanced signal at the sensor.

1.5. IMAGE CONTENT AND IMPERFECTIONS

1.5.1. Imperfections

The combined imaging and lighting process applies to a scene in three dimensions, and over time, where each point is characterized, in a wavelength-dependent fashion, by its electromagnetic emission, transmission, diffraction, refraction, diffusion, reflection, and absorption.

The wavelengths of electromagnetic emission are given in Figure 7, and typical scene lighting data are presented in Table 3. Specular and diffuse illumination are depicted in Figure 8. The minimum illumination levels as a function of object size are given in Table 4.

The imaging sensor (planar) loses parts of this information, for the following basic reasons:

1. The imaging sensor carries out a conical projection of the scene from three into two dimensions (nonreversible).

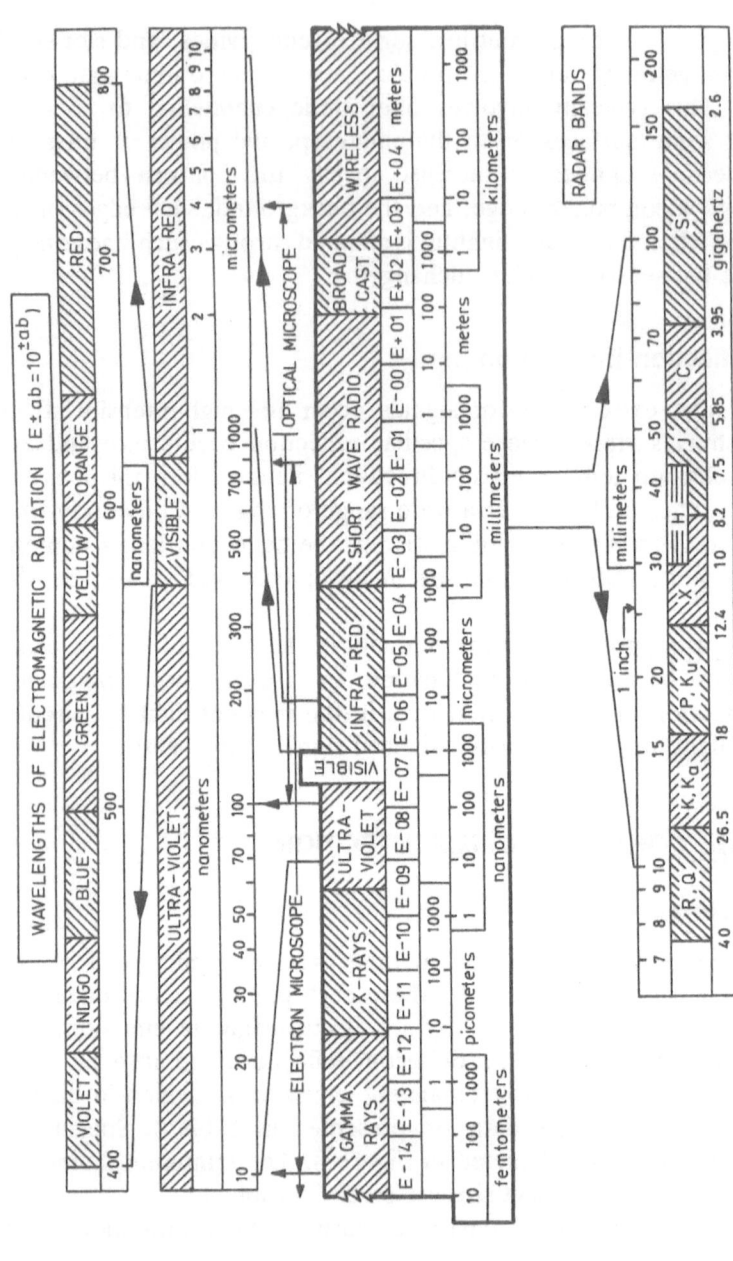

Figure 7. Wavelengths of electromagnetic radiation; $X \cdot E$ means $X \cdot [10 \cdot \exp(\pm ab)]$.

Table 3. Typical Scene Illuminances and Sensor Ranges

Situation	Illuminance (lux)	Sensor	
Direct sunlight	$(1–1.3) \times 10^5$		
Full daylight	$(1–2) \times 10^4$		
Overcast sky	10^3		
Very dark day	10^2		Human eye
Twilight	$1–10^2$ }	Vidicon	
Deep twilight	1		
Full moon	10^{-1}	Orthicon	
Quarter moon	10^{-2}	Isocon	
Moonless, clear	10^{-3}	SIT	
Moonless, overcast	10^{-4}	Image intensifiers	

2. The imaging sensor integrates the electromagnetic energies listed above over time, irrespective of scene dynamics.
3. Each point in the image plane is the radiometric mix of information originating in several points and directions in the three-dimensional scene.
4. Each point in the image plane is the geometrical mix of information originating in several planes intersecting with the line of sight.
5. The imaging optics generate geometrical and signal anomalies— most significantly, chromatic aberration, spherical aberration, coma, and distortion (see Figure 9).

An image sequence delivered by the sensor will not be triggered, or in general synchronized, by events in the scene, unless other systems are used.

Finally, the image, if digitized into a $N \times M$ pixel array, with a given signal quantization scheme at each pixel (see Algorithm Quant-1), will be furthermore degraded owing to

1. Spatial/grid sampling.
2. Signal quantization.
3. Spectral sampling.

1.5.2. Lighting Environments

Visible light is the name given to the part of the electromagnetic spectrum between 380–400 and 750–780 nm because the human eye can detect it. The maxima of the eye's sensitivity are at 556 nm in daylight

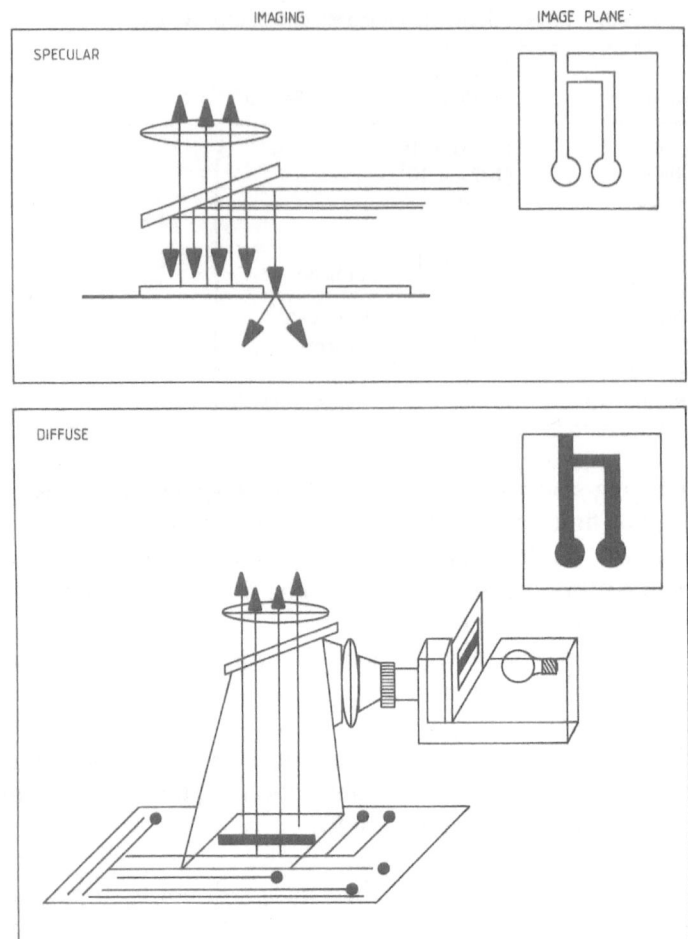

Figure 8.   Setups and effects of specular (top) and diffuse (bottom) illumination.

(photopic vision) and at 507 nm at night (scotopic vision). The wave-length intervals coresponding to the dominant colors are (see Figure 7) as follows:

380–436 nm: violet
436–495 nm: blue
495–566 nm: green
566–589 nm: yellow
589–627 nm: orange
627–780 nm: red

Table 4. Minimum Illumination Levels According to Object Feature Size and Object Contrast

Object feature size class (and angular size)	D/d interval	Minimum illumination intervals (in lux)		
		High contrast	Medium contrast	Low contrast
Very small	3200	4000	10,000	40,000
($\alpha < 1'05''$)		1800	5,000	18,000
Very fine	2450	850	2,500	8,500
($1'05'' < \alpha < 1'25''$)				
Fine	1900	450	1,200	4,500
($1'25'' < \alpha < 1'50''$)				
Fairly fine	1500	200	600	2,000
($1'50'' < \alpha < 2'20''$)				
Medium	1150	90	280	900
($2'20'' < \alpha < 3'$)				
Coarse	850	40	130	400
($3' < \alpha < 4'05''$)				

For emitted light, the photometric variable is the luminance, with the candela as the unit thereof.

For received light, the photometric variable is the illumination, with the lux as the unit thereof. An illuminated surface may be emissive by reflection, in which case it too emits candelas.

Light also has color attributes in human vision, because the photosensitive cell inputs are recombined in the nervous system. A color stimulus has essentially three such attributes (see Figure 10):

1. Intensity, which is the psychophysiological luminance.
2. Hue, which gives the color value and is related to the dominant spectral wavelength.
3. Saturation, which gives the proportion of the pure chromatic sensation in the total sensation; it is approximately equivalent to the chromatic purity, i.e., the proportion (between 0 and 1) of white added to the fundamental wavelength; the grade 0 corresponds to an achromatic stimulus roughly equivalent to white.

Physiology has shown that any color sensation can be synthesized from three fundamental colors, yielding the color-triangle representation:

$$s = x\mathrm{A} + y\mathrm{B} + z\mathrm{C} \qquad \text{with } x + y + z = 1$$

where A represents red, B green, and C blue, s is the colored light source, and x, y, and z are the luminance contributions of each

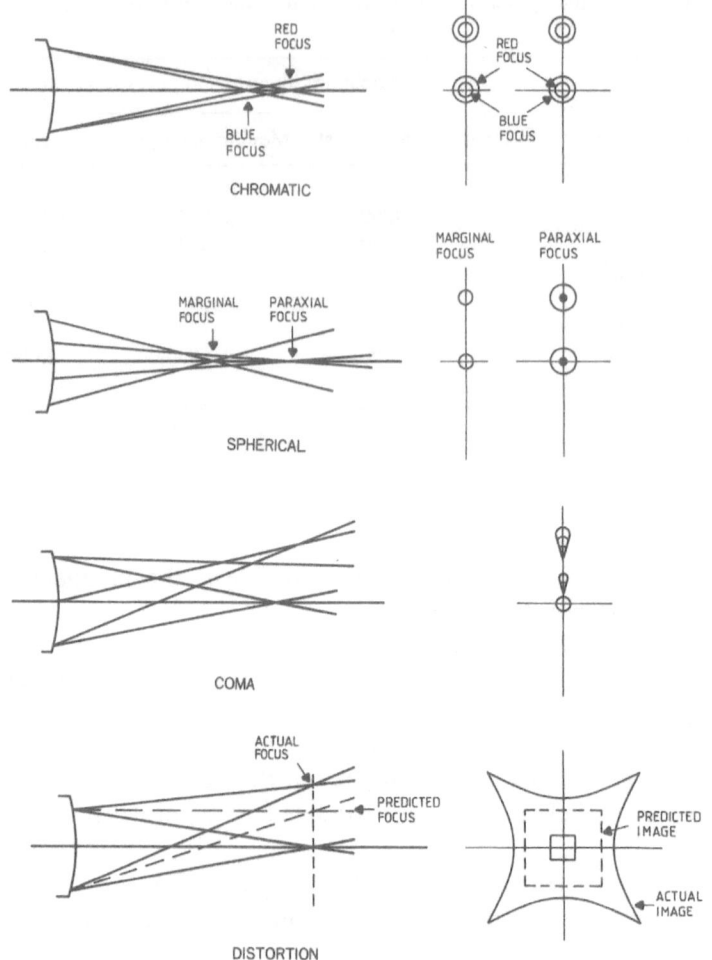

Figure 9. Aberrations due to optics: chromatic aberrration, spherical aberration, coma, and distortion.

fundamental color. The trichromatic coordinates (x, y, z) can be measured by a colorimeter (yielding x, y) or can be computed from the emission spectrum of the light source (measured by a spectroradiometer).

1.5.3. Lighting Design

The lighting design results from an ergonomic analysis to determine exactly which lighting characteristics are needed by the vision system

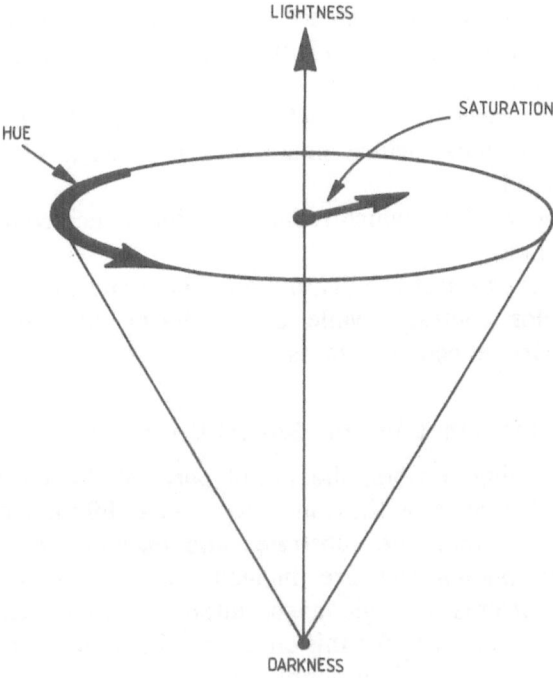

Figure 10. Color perception attributes: Hue, lightness/intensity, and saturation.

application. The following factors are considered in the lighting design:

1. *Angular object size* is the aperture angle α of the cone placed at the sensor location, which encompasses the scene to be viewed:

$$\alpha = d/D \quad (\text{rad})$$

where d is the scene diameter (m), and D is the distance (m) between the sensor and the scene.

2. *Luminance contrast* is the ratio of the scene luminance L_0 to the background luminance L_F:

$$C = (L_F - L_0)/L_F$$

The higher the contrast, C, the better is the visual acuity.

3. *Background luminance, L_F*: not only affects the contrast, C, but also the adaptation level of the eye sensor; in the eye, either the photopic system, with cone detectors, or the scotopic system, with rod detectors, is enabled.

4. *Color contrast* is the distance between the trichromatic coordinate values of the scene and those from the background. Color contrast compensates for lack of luminance contrast. Hue distance is especially useful, whereas saturation distance is not.
5. *Disability glare,* which neutralizes the vision sensor, must be eliminated.
6. *Discomfort glare,* which reduces the luminance contrast, must be eliminated.
7. *Color spectrum* of the lighting must be matched so as to enhance the color contrast, while also reducing the effects on it of secondary reflective surfaces.

1.5.4. Image Compensation for Spatial Emissivity

For all imaging sensors, the useful parts of the electronic device under test (DUT), such as the conductors, have different emissivities at some wavelengths than the substrates and background. Furthermore, when dynamic phenomena are imaged (dynamic temperature rise, photoinduced currents for logic maps, differential contrast), it is particularly crucial to compensate for this emissivity difference. Very few bodies behave as blackbodies, and all nonblackbodies reflect energy.

Therefore, the following two-step setup procedure is required:

1. Record the radiance image of the DUT with effective radiance expressed as $mW/cm^2 \cdot sr$, which is the effective blackbody radiance in the spectral range of the sensor; this requires temperature readings and solution of the Planck equations.
2. Record the radiance image of the unpowered DUT, at a known temperature, and calculate from this radiance image and that recorded in step 1 the emissivity at each pixel, also called the emissivity image.

Thereafter, using the radiance image recorded in step 2, any radiance image of the same type of DUT can be converted to a temperature image or induced current image, and defect detection carried out on the latter.

1.6. DESIGN CHOICES FOR THE VISION SYSTEM SPECIFICATIONS

The design choices and specifications for a vision system result from a delicate balance between systems performance requirements and system architecture. Whereas there are many important details to be sorted out at the design stage, it is crucial never to forget the essential elements of the choices, which are sometimes forgotten by designers and are therefore listed below.

1.6.1. Performance Requirements (R)

The performance requirements to be specified include:

R0: List of defect types to be detected, with their detectability thresholds in terms of size, area, shape, depth, and central emissivity wavelength.

R1: Nondetection probability (i.e., probability of not detecting a defect), over the range R0.

R2: False alarm probability (i.e., probability of claiming a good item to be defective), over the range R0.

R3: Maximum and average defect detection and classification delay (excluding handling, transport, and mechanical alignment).

R4: Like R3, but including handling, transport, and mechanical alignment.

R5: Access (or not) and, if yes, access time to CAD or layout data (see Algorithm CAD-1).

R6: Access (or not) and, if yes, access time to on-line process data.

R7: Reconfiguration and setup time for another type of device.

R8: Defect detection or classification, and if classification is required, into how many classes of defects (see Chapter 20).

1.6.2. Architecture (A)

The major elements in the design of the system architecture are:

A0: Type of sensor (see Section 1.1).

A1: Sensing principle, as related to the physics of the inspection task (see Table 2).

A2: Sensor scanning geometry and speed (see Section 1.3).

A3: Size and resolution in each area of interest (see Algorithm AOI-1).

A4: Informational content in the image, for each defect type, defined as the ratio of the number of defect image pixels to the background pixel number in the area of interest.

A5: Image data bus hardware, speed, and protocol.

A6: Architecture of the image feature extraction and evaluation processor (see Chapter 19).

A7: Architecture and speed of the image processor, and degree of parallelism thereof.

A8; Real-time operating system (O.S.) and language for image analyses, including symbolic processing (see Chapter 8).

A9: Real-time O.S., interrupt controller, and programming language for image acquisition, image preprocessing, and task scheduling.

A10: Bandwidth of link between sensor and image frame store or image buffers.

A11: Bandwidth and addressing scheme between A9 and A7.

A12: Bandwidth and addressing scheme between A7 and A6.

A13: Storage capacities needed for reference images, knowledge bases, and CAD data files.

1.6.3. Vision System Design and Test Aids

Present design methods for vision systems are very labor-intensive and, moreover, demand a very high level of skill on the part of a multidisciplinary team. Therefore, computer-aided engineering tools are a must in this area too, focusing on the following subsystems:

1. Knowledge-based system to advise about lighting and optical configuration (see Chapter 8).
2. Knowledge-based system to assemble and configure together various algorithms and test them out on an interactive image analysis system; this corresponds, for example, to all the tables entitled "Basic Steps" in this book.
3. Software for configuring the vision and other processor boards.
4. Fast image processor, with image archival facilities, to emulate the target processor and collect data under real conditions.
5. Software for conversion of CAD, drill tape, and schematic drawing files into computer graphic layouts (see Chapter 21).

Concerning test aids to be used during vision system setup, several are needed to check accuracy and repeatability, that is, to check that the measurements are taken within a given tolerance every time. The tests include:

• Camera and lens calibration (see Algorithm Calib-1).
• Dot reproducibility.

- Scanner/sensor reproducibility.
- Positional accuracy.
- Autoregistration reproducibility (see Chapter 15).
- Data base-to-board registration verification.
- "Golden board" or reference mask reproducibility (see Algorithm Templ-1).

1.7. CALCULATION OF THE INSPECTION YIELD

This section gives a few calculation rules to help in the evaluation of an inspection procedure (see Section 1.6.1).

1.7.1. Defect Detection Probability

If the defects are randomly distributed in size, shape, and location, the probability of detection, P, is best treated as a Poisson distribution with

$$P = p^r(1 - p)^{(n-r)}(n!)/(r!)(n - r)!$$

where n is the defect sample size, r is the number of defects detected in the sample of size n, p is the true fraction of defects in the population of items, and (r/n) is the apparent probability of detection.

Example: $p = 0.20$ $n = 5$

r	P (%)	r/n (%)
0	32.77	0
1	40.95	20
2	20.48	40
3	5.12	60
4	0.64	80
5	0.032	100

1.7.2. Defect Probability from Area and Image Density

It is common to express p, the true fraction of defects in the population of items, by assuming an exponential relation to defect area and density of defects in the image:

$$p = \exp(-a \cdot D)$$

Table 5. Investment Return Model: Example of Hybrid Circuits with Screen Printer[a]

	Current visual inspection	Automated optical inspection	
		Hand-fed	Automatic
Production:			
Work day (min)	480	480	
Time per set-up (min)	40	20	
Setups per day	6	12	
Available production time (min)	240	240	
Per-piece information:			
Print time (s)	5	5	
Load, inspection, unload time (s)	60	15	
(4″ × 5″, 1-mil res., nominal complexity)			
Average repair time (s)	15	15	
Probability of repair	5.0%	7.5%	
Total print/inspection/repair time (s)	66	21	
Less: Concurrent printing	−5	−5	
Average print/inspection/repair time (s)	61	16	
Prints per day:			
Number of stations	1	1	
Total prints per day	237	893	
Days per year	240	240	
Annual product yield:			
Annual production	56,889	214,326	
Prints per substrate (inspected)	27	27	
Annual throughput (substrates)	2,107	7,938	7,938
Average yield percentage	50%	50%	55%
Shippable product	1,054	3,969	4,366
Return on investment:			
Annual production improvement		2,915	3,312
Substrate value		$500.00	$500.00
Increased annual shipments with automatic optical inspection		$1,457,747	$1,656,197
Gross margin percentage		15%	15%
Pretax profit before depreciation		$218,662	$248,430
Less: Annual depreciation expense[b]		$(26,600)	$(26,600)
Estimated annual pretax profit increase		$192,062	$221,830
Payback period (months)		7.87	6.82

[a] Source: Photo Research Vision Systems.
[b] Investment analysis:

Total system cost (estimated)	$140,000
Investment tax credit	−14,000
Net investment	$126,000
Depreciable basis: 95.0% of $140,000	= $133,000
Average annual depreciation $133,000/5 =	$26,600

where a is the area of a defect in the image, and D is the image density of such defects in the image. The density D can be either the initial defect density before any repair or the residual density after repair.

Example: $a = 0.30$ $\quad$ $D = 2.5$ $\quad$ $p = 0.47$

1.8. TOTAL INSPECTION COSTS

If an n-step inspection procedure is used, the total inspection cost, T, is

$$T = \sum_{i=1}^{n} \sum_{j=i}^{n} \frac{C_i}{\pi(1 - p_j)}$$

where C_i is the cost of inspection at step i, p_i is the true fraction of defects in the population tested at step i, and $(1 - p_i)$ is the yield of that step.

The total inspection cost is, of course, test/inspection sequence dependent. See Table 5 for a complete return of investment model, in a specific case.

Examples:

1. Inspection sequence by ascending yield:

$$\left.\begin{array}{ll} C_1 = 40 & p_1 = 0.5 \\ C_2 = 20 & p_2 = 0.4 \\ C_3 = 10 & p_3 = 0.1 \end{array}\right\} T = 196.3$$

2. Inspection sequence by ascending costs:

$$\left.\begin{array}{ll} C_1 = 10 & p_1 = 0.1 \\ C_2 = 20 & p_2 = 0.4 \\ C_3 = 40 & p_3 = 0.5 \end{array}\right\} T = 183.6$$

3. Minimum-cost inspection sequence:

$$\left.\begin{array}{ll} C_1 = 20 & p_1 = 0.4 \\ C_2 = 10 & p_2 = 0.1 \\ C_3 = 40 & p_3 = 0.5 \end{array}\right\} T = 176.2$$

Chapter 2

Imaging Microscopes for Microelectronics

Various types of microscopes are used in microelectronics. These differ in the illumination fields used (vertical, oblique, darkfield, interferential, differential), light sources, iris and diaphragms, and filters (see Tables 6 and 7).

2.1. OPTICAL MICROSCOPE ATTRIBUTES FOR MICROELECTRONICS

2.1.1. Microscope Lighting

For the microscope illumination system, the choice is among the following:

1. Illumination spectra, wideband or single band (laser).
2. An epi-illumination, incorporating either brightfield–darkfield (BD) or differential interference contrast (DIC) objectives, or both. A rotating polarizer with analyzer is also useful.
3. A fluorescence illuminator, which causes photoresist to fluoresce; it holds a mercury bulb with UV emission steered by UV filter cubes.

Some selection guidelines are given below for various classes of

Table 6. Microscopes for Microelectronics

Microscope type[a]	Magnification factor	Use for IC inspection[b]	Reference(s)
Stereo	0.7–160	+	
Split field	7–40	++	
Metallurgical	25–1500	+++	
Polarized light	50–700	++	
Interferometric/Nomarski	25–500	+	
Metrological	20–1000	+++	
Fluorescence	30–200	+	2
Infrared	50–200	+	3,4
Ultrasound	30–600	++	5
Polarized infrared	50–150	+	6
Optical section	50–200	++	

[a] The best resolution with an optical microscope is, however, about 0.25 μm with fixed magnification of maximum $\times 2000$.
[b] The usefulness for IC inspection is graded from + (lowest) to +++ (highest).

applications:

1. *General inspection:* Brightfield illumination is required to check for residual resist, scumming, displacement of polymers, completeness of the strip/pattern alignment and integrity, and film thickness. To check for substrate defects, etching, residual metal, polysilicon, or residual oxide or nitride ions, darkfield illumination is used with BD lenses. To show the surface level in checking for surface step coverage, to obtain clear definition, and to check the step slope, DIC illumination is useful.

Table 7. Some Typical Optical/Digital Parameters Used in Video Inspection Systems[a]

Optical field of view	Approximate magnification	Approximate depth of focus	Digital resolution
3″	0.25×	0.200″	0.006″
1″	0.75×	0.100″	0.112″
0.5″	1.5×	0.050″	0.001″
0.1″	7.5×	0.010″	0.0002″
0.05″	15×	0.005″	0.0001″
0.01″	75×	0.001″	0.00002″
0.005″	150×	0.0002″	0.00001″
0.002″	375×	0.0001″	0.000004″

[a] Ref. 7.

2. *Process control:* To check alignment and profiles, the combination BD/DIC with ×150 magnification is common.
3. *Mask inspection:* Transmitted light microscopes (trinocular) are usually required (see Chapter 7).
4. *Line inspection:* A binocular microscope is usually employed (see Chapter 3).
5. *Critical dimensions:* Can make necessary special measuring apparatus, and thus lead to a quadrocular body tube (see Chapter 3).
6. *Photoresist inspection:* Fluorescent illumination is often required to reveal resist before or after ion implantation (see Chapter 6).
7. *Wafer inspection:* DIC illumination reveals the crystal lattice structure (see Chapter 6).

2.1.2. Confocal Imaging

Confocal imaging involves placing a small pinhole in front of the detector, so that it behaves as a point detector[8] (Figure 11). The image of a point object is 37% sharper in confocal imaging. This principle holds for optical as well as laser scanning microscopes.

The confocal arrangement also results in an optical sectioning property. Light scattered by parts of the object away from the focal plane is defocused at the detector pinhole and hence is detected weakly. Out-of-focus parts just do not contribute to the image and are not

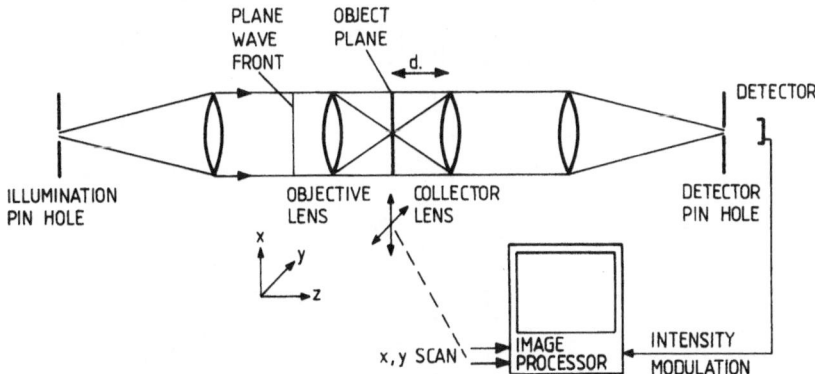

Figure 11. Confocal microscopy. A plane light wave is brought to a focal point by the objective lens. The collector lens is so positioned that the back projected image of the detector pinhole exactly coincides with this focal point. The object is *x–y* scanned through the common focal point, modulating the light intensity on the detector. The image processor is synchronized with the object scan.

blurred. By scanning the object in the z-axial direction, it is possible to build up a series of image slices. This gives range mapping (see Section 2.3.3).

If the image slices are summed up digitally, a greatly extended depth of focus can also be obtained. If not, pinhole focusing provides an unambiguous means of focusing at a point the size of the resolution cell, instead of relying on contrast variations.

An internal interferometer must be used to set the scanning distances and, together with the digital sample time, determines the size of each pixel in the image. This is a useful feature for device metrology.

2.2. ELECTRON BEAM INSPECTION OF ICs

2.2.1. Principle of Scanning Electron Microscopy (SEM)

An electron beam (e-beam) may replace light or mechanical probing. An e-beam can be focused down to 3 nm, and beam scanning is

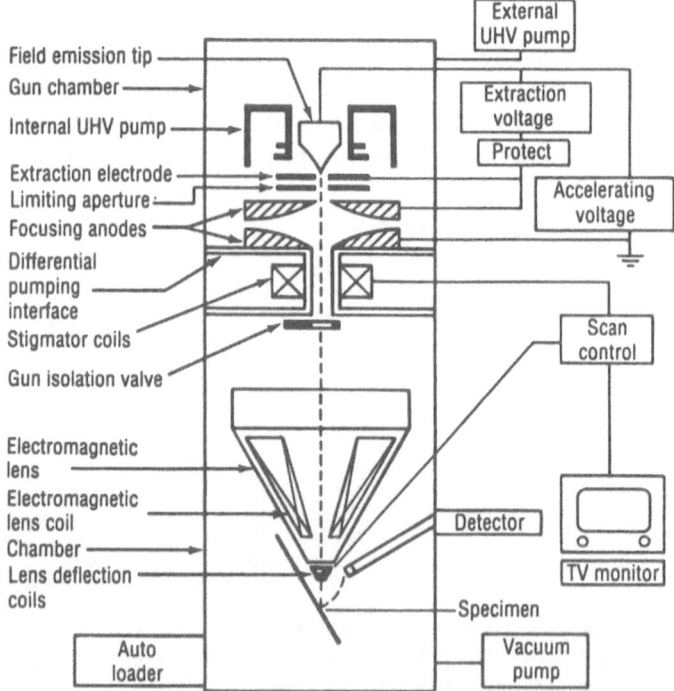

Figure 12. Scanning electron microscope (SEM) with field emission (FE) source and focus control.

achieved by the use of electromagnetic deflection coils (Figure 12). At the surface, the *e*-beam generates low-energy secondary electrons (SE), in a process dependent on material characteristics, circuit voltages, and geometry. The SEs are picked up by detectors, and this brightness modulates a cathode-ray tube in which the beam is deflected in synchronism with the SEM. This allows also for the capability of imaging voltage distributions on the chip and of measuring signals through passivation and oxide layers.

2.2.2. Voltage Resolution

The SEM voltage resolution is limited by the shot noise in the primary electron beam: this shot noise decreases with increasing beam current. Since the beam current affects the spot size, *d*, the voltage resolution, V_{min}, is related to the spatial resolution by

$$V_{min} = C \cdot (Df)^{1/2}/d$$

where C is a constant of the equipment, and Df is the system bandwidth. Typically, for $d = 0.1\,\mu m$, Df $= 100\,kHz$ and $V_{min} = 50\,mV$.

2.2.3. Stroboscopy

Leaving the phase between the *e*-beam pulse and the IC signal fixed while scanning the IC surface with the beam generates micrographs of a frozen logic state. Subsequent shifts of the phase yield stroboscopic micrographs of different logic states[9] (see Section 2.5).

2.2.4. Dynamic Fault Imaging (DFI) for Timing Problem Analysis

The basic steps in dynamic fault imaging are listed in Table 8. The IC is exercised by a pattern generator, and stroboscopic images are acquired from the device under test (Figure 13). The images of the corresponding logic states are subtracted (Algorithm Subtr-1). Image enhancement is carried out through table lookup (Algorithm LUT-1). Doing this for a number of subsequent logic states by image sequence analysis reveals the first occurrence of a failure and its subsequent spreading until the bond pads are reached.

The logic state images can be synchronously subtracted from those of a reference device or from the logic state image of the same device in the previous logical state. This activity mapping provides information about

Table 8.   Basic Steps in Dynamic Fault Imaging (DFI)

Sensor: Scintillator tube.
Attitude: Top-down view.
Illumination: SEM, synchronized with IC state sequence, or alternatively, LSM.

0.1 Setup of logic levels of IC tester, primary SEM beam current, blanker pulse width, and electric field induced from adjacent pixels.
0.2 Setup of exponential moving average of image frames by video mixing.
0.3 Setup of counter registers telling how often each pixel in the image was sampled.
1.0 LUT-1: LUT-based image enhancement.
2.0 Subtr-1: absolute difference of two successive images.
3.0 Edge-1: filtering-based edge detection in step 1.
4.0 Pseudo-1: pseudo-color coding of latches.
5.0 Loop on step 1 until end of test input sequence.
6.0 Movie display of the image sequence in step 4.
7.0 Comparison with IC simulation model, e.g., SPICE.

Special aspects:
• 700-eV–2-keV e-beam, with 10–20-nm penetration, giving an average EBIC current of 350–500 pa.
• Voltage resolution of 0.1–0.3 V.
• e-Beam pulse width of 2 ns–150 ms.

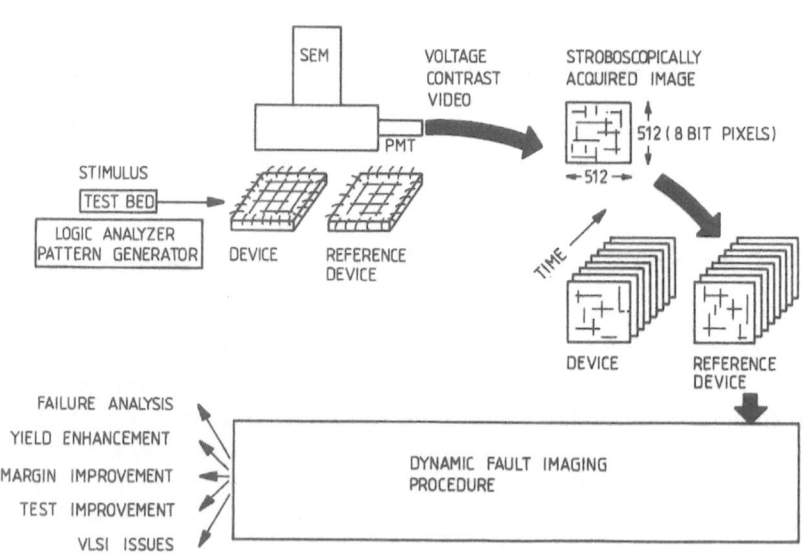

Figure 13.   Dynamic fault imaging (DFI) with SEM.

test pattern activities, stress tests, and signal traffic. Alternative approaches use pulsed IR microscopy[3] (Section 2.4) and laser scan stroboscopic imaging of inversion layers [10,11] (Section 2.5). See also Refs. 9 and 12 for further details on DFI by SEM.

2.2.5. Thermal Gradient Imaging of Junctions

The principle of thermal gradient imaging of junctions is to combine the SEM induced current (EBIC) and the characteristic change in the forward voltage drop of junctions with temperature to produce a temperature reading at the spot probed by the beam. This requires, however, that the calibration curve of the junction be determined initially. The SEM parameters used are the e-beam current and acceleration potential, as well as the spot size. The junction potential measurement requires a very high input impedance amplifier. See also Ref. 13.

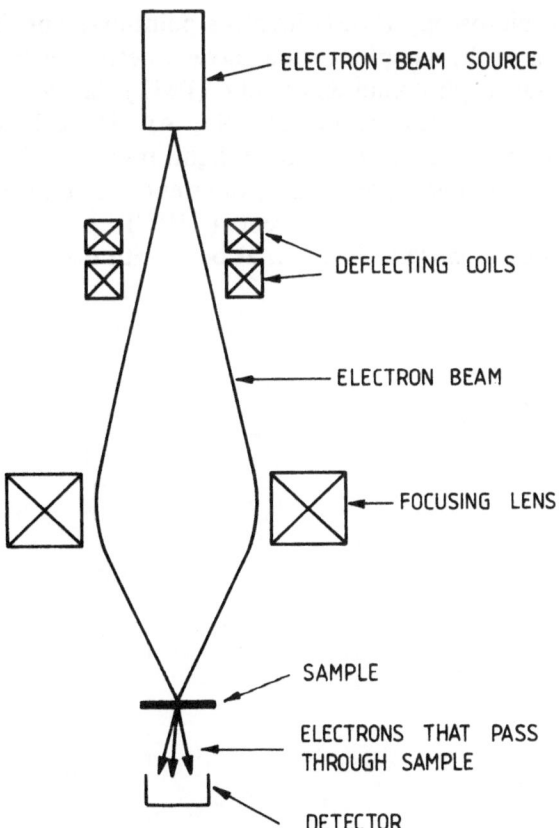

ELECTRON‐BEAM SOURCE

DEFLECTING COILS

ELECTRON BEAM

FOCUSING LENS

SAMPLE

ELECTRONS THAT PASS THROUGH SAMPLE

DETECTOR

Figure 14. Scanning transmission electron microscopy (STEM).

2.2.6. Scanning Transmission Electron Microscopy (STEM) for Cross-Sectional Analysis

As an alternative to SEM, which suffers from sample preparation and contamination problems, scanning transmission electron microscopy (STEM) can be employed (Figure 14). STEM scans the wafer, die, or circuit. The sample must be thin enough to allow the e-beam to pass through it with high enough attenuation to provide sufficient contrast, but not so high that too few electrons reach the imaging detector. STEM is superior for cross-sectional studies, because the contrast results from passing the beam through the slice's bulk.

2.3. LASER SCAN MICROSCOPY

2.3.1. Principle of Laser Scan Microscopy (LSM)

Laser scan microscopy (LSM) involves point-by-point line raster or random scanning of the sample, with a galvanometric beam deflection, a beam splitter, and a photomultiplier tube (PMT) detector[14-17] (Figures 15–17). The laser sources include He–Ne, Ar, He–Cd, and excimer lasers. The contrast types include incident light, transmitted light, crossed polarizers, phase contrast fluorescence, differential interference contrast (DIC), and optical beam-induced currents (OBIC).

The resolution is close to $0.25~\mu m$ at best, with magnifications up to

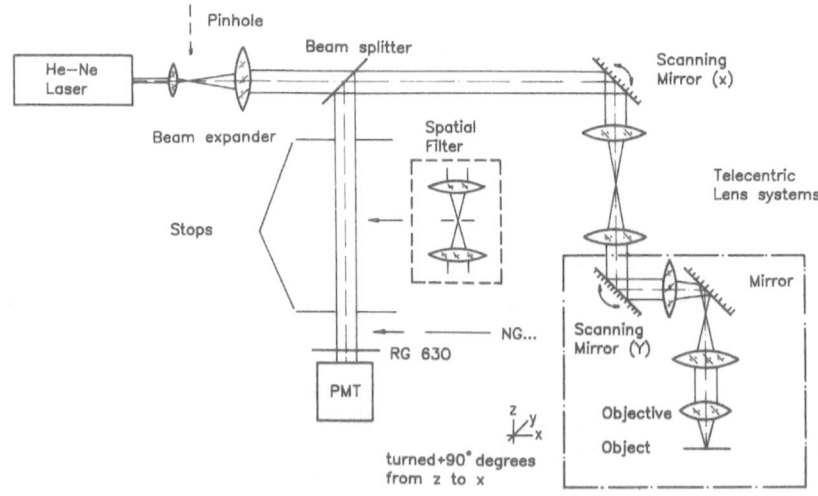

Figure 15. Laser scan microscope (LSM) and beam path.[14] (Courtesy Carl Zeiss.)

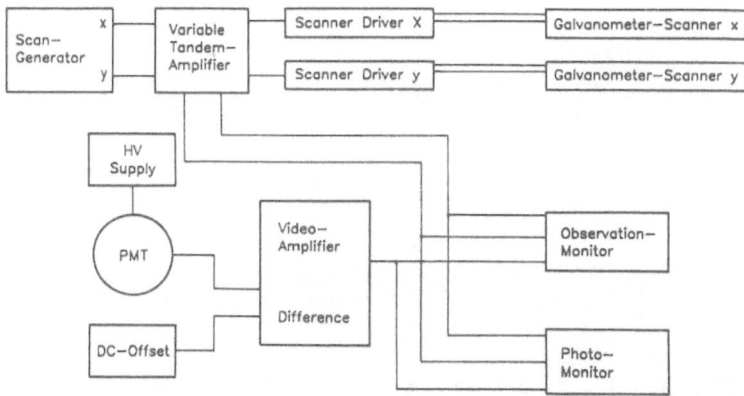

Figure 16. Laser scan microscope (LSM) electronics. (Courtesy Carl Zeiss.)

×16,000 with, for example, scanning argon ion confocal laser micro-scopes. The digitized video signal can be processed as a 512 × 512 or 1024 × 1024 pixel image, with a frame repetition period of 2–64 s (Figure 18).

2.3.2. Laser Spot Size

The laser spot size, $d(z)$, on the sample after beam focusing by a lens must be minimum, meaning that the beam enlargement must be controlled.

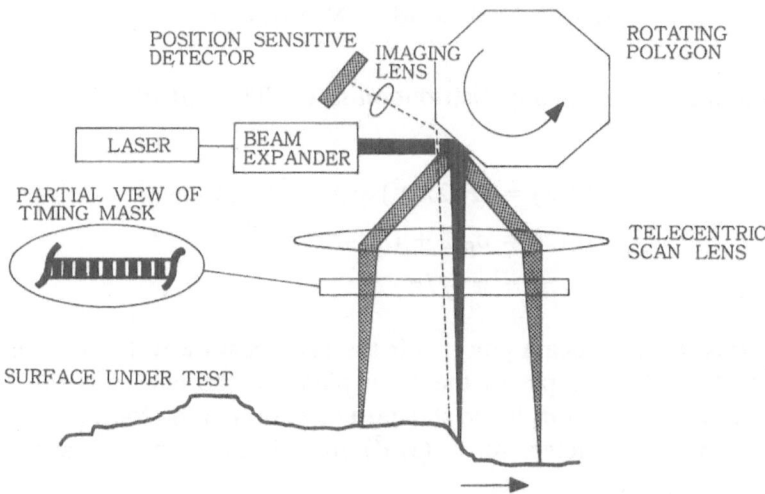

Figure 17. Raster scanning for 3-D LSM.

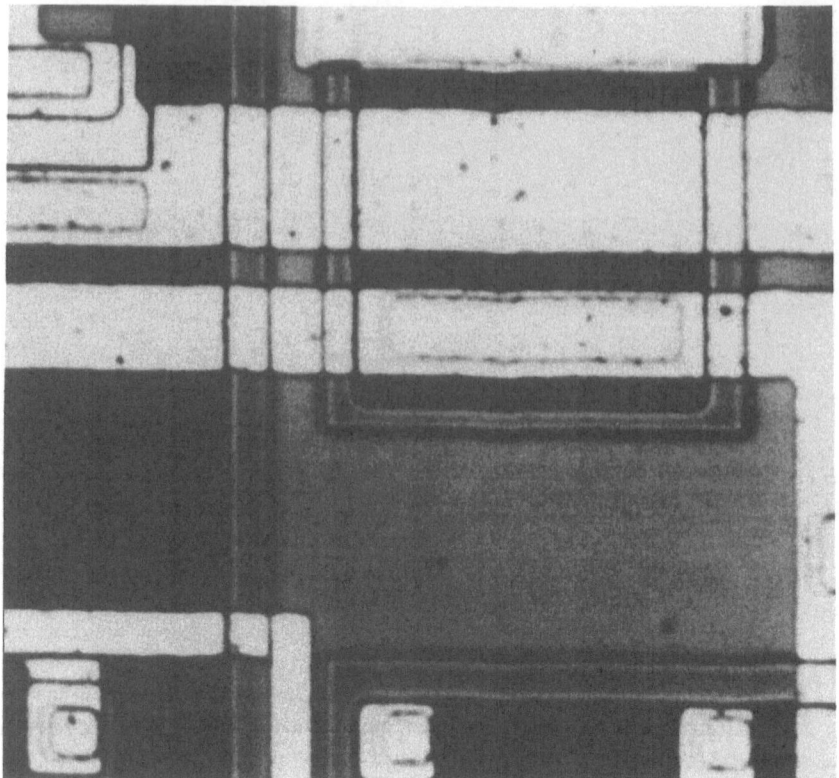

Figure 18.   Incident light LSM image of an IC.

The laser beam energy distribution in the TEM $(0, 0)$ mode is given by[18]

$$I(r, z) = (P/2\pi\sigma^2)\exp(-r^2/2\sigma^2)$$
$$\sigma(z) = \sigma_0[1 + (lz/4\pi\sigma_0^2)^2]^{1/2}$$
$$\sigma_0 = \pi \cdot f/\pi \cdot d$$

where P is the laser beam power, l is the laser beam wavelength, z is the distance from the sample to the focal plane of the lens, f is the focal length of the lens, and σ_0 is the standard deviation in the focal plane. The resulting beam diameter, for a $(1/e^2)$ ratio between border and peak energy, is given by

$$d = 4\sigma(z) \qquad d_0 = 4\sigma_0$$

If a beam expansion ratio ε is assumed (e.g., $\varepsilon = 0.1$), then

$$(d - d_0)/d_0 = [\sigma(z) - \sigma_0]/\sigma_0 = \varepsilon$$

The above equations can be solved for z and d.

2.3.3. Laser Scanning Tomography (LST)

The combined use of confocal LSM, autofocusing, short-wavelength lasers, and image processing allows for the generation of a range image showing variations in specimen height[19] (Figure 19). This method consists of labeling each pixel with the digitized output from the confocal

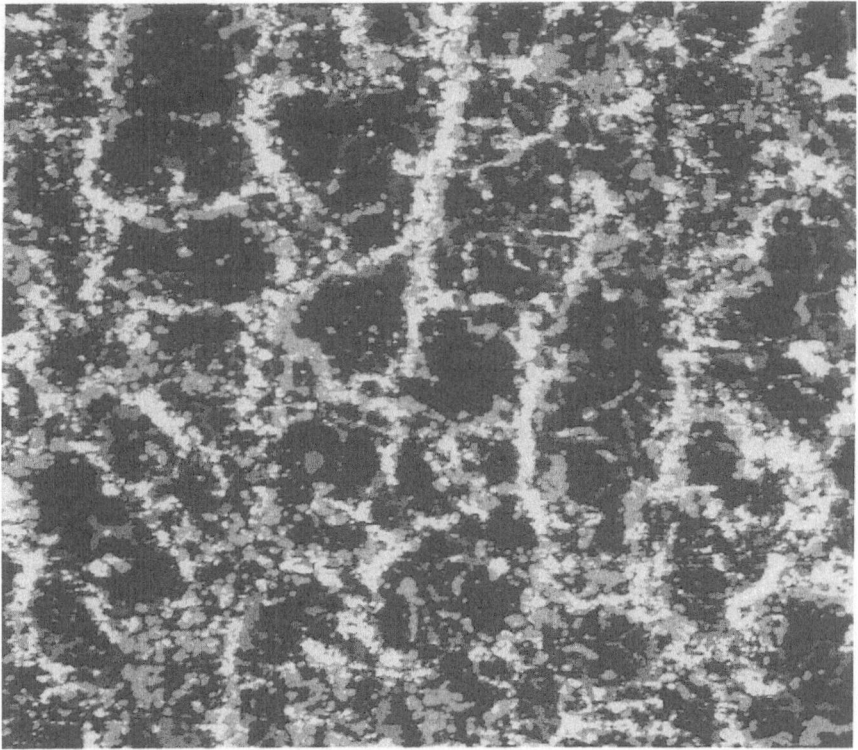

Figure 19. Image obtained by laser scanning tomography (LST). An IR real laser beam was introduced into the wafer sample, through a flat coplanar lateral section; the scattered beam due to imperfections in the material was then imaged to reconstruct a 2-D image after moving the sample; a pattern of cells, streamers, point precipitates, and linear dislocations is revealed with a 10-μm resolution.

autofocus, to find the maximum output, applying a pseudo-color coding (Algorithm Pseudo-1), and overlaying the image slices. The accuracy and speed are low, but it is still a useful technique, and the wavelength can be selected through the use of, for example, excimer lasers.

2.3.4. Applications

The LSM gives high-contrast imagery, with addressable point locations, as is useful for OBIC[15,16] (Section 2.5); dynamic fault imaging (DFI), e.g., for latch-up effects[15,17]; hot spots[17]; and diffusion layers.[17]

The defect detection rates, for each class of defect, must be established through experimental work, using defective ICs with characterized defect types. This means that gaining access to such defective ICs (not available generally, except from IC manufacturers) is advisable.

One can indeed generate a voltage contrast image (not absolute voltages, however) as a variant of the current density image. This does require the addition of some signal generators and synchronizing trigger circuits; sensing time is about three times longer.

2.4. PULSED IR MICROSCOPY

2.4.1. Dynamic Latch-Up Imaging

Infrared microscopes have the unique property of pinpointing forward-biased junctions on silicon devices. Such forward-biased junctions in silicon emit a tiny amount of radiation by recombining excess electrons and holes; this recombination radiation is emitted at a wavelength of $1.1\,\mu m$[20] (Figure 20). The IR microscope reveals the

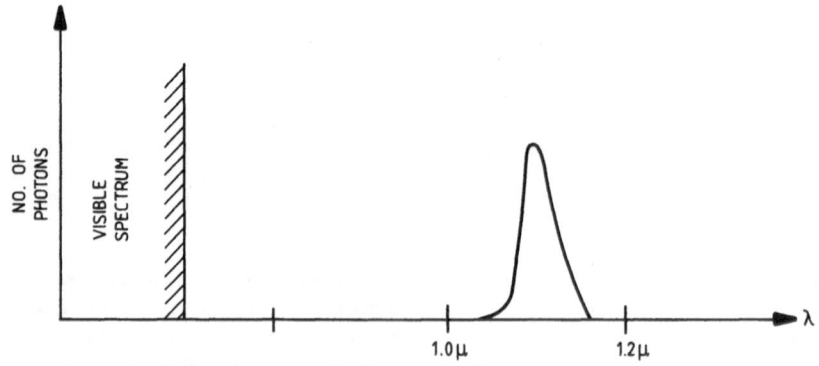

Figure 20. Relative emission spectrum of silicon.[20]

junctions, but only image processing of a sequence of such images by dynamic fault imaging (see Section 2.2.4) allows analysis of latch-up by revealing how it spreads into the IC.[3] Control of the dynamic fault imaging is achieved by pulsing the power supply.

2.4.2. Static Substrate and Circuit Analysis

The optical properties of III–V compounds, garnets, and ferrites are such that they are opaque in the 450–700-nm visible region and quite transparent in the 900–1200-nm near-IR region. Silicon becomes transparent beyond 1150 nm. Moreover, thin metal layers are often transparent in the near-IR region. Infrared image converter tubes operate until about 1200 nm, although a filter must be used to compensate for lack of lens color correction.

The most useful applications[4,21] of IR microscopy imaging therefore are the examination of metallizations on semiconductors, of dislocations and subsurface defects, and of die bonds, through the reflected IR light,[22] and metallization/silicide inspection.

2.5. IMAGING OF PHOTOINDUCED CURRENTS

Photoinduced currents are essentially generated (Figure 21) as optical beam-induced currents (OBIC), by a laser scan microscope (LSM), or as electron beam-induced currents (EBIC), by a scanning electron microscope (SEM) or equivalent.

OBIC probing requires a normal operational environment (clean room atmosphere), as opposed to SEM or EBIC probing, which must take place in vacuum.

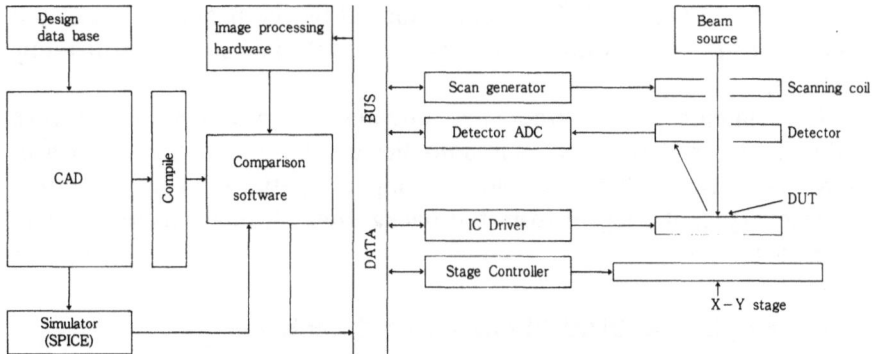

Figure 21. Principles of OBIC or EBIC inspection testing. For OBIC, the source is a laser scanning device (LSM); for EBIC, it is a scanning electron microscope (SEM).

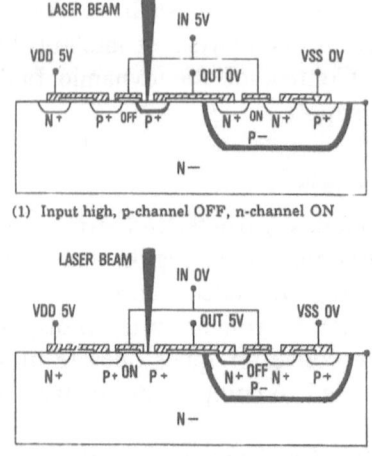

(1) Input high, p-channel OFF, n-channel ON

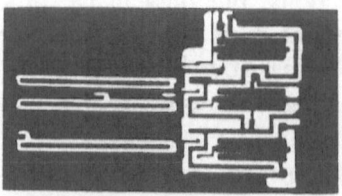

(1) Input high, p-channel OFF, n-channel ON

(2) Input low, p-channel ON, n-channel OFF

(2) Input low, p-channel ON, n-channel OFF

Figure 22. Laser-based scanning OBIC imaging of a CMOS inverter: (1) input high, *p*-channel OFF, *n*-channel ON: (2) input low, *p*-channel ON, *n*-channel OFF.

2.5.1. Optical Beam-Induced Current (OBIC) Measurements

The OBIC technique uses a focused scanning laser beam (LSM) to produce electron–hole pairs. Such pairs are produced when the laser wavelength is such that the photon energy is greater than the bandgap of the target material (Figure 22). In indirect bandgap semiconductors such as silicon, the depth of penetration of the laser beam depends on the wavelength. The probe beam can be modulated, and a bias voltage applied to certain sample leads. Proper signal amplification and grounding are essential.

When the laser beam is scanned across the sample, a short-circuit current may be observed if carriers produced by the beam are separated from built-in fields within the sample (defects, strain, protruding junctions).

Experiments in the frequency domain, using a continuous-wave modulated laser, a vector lock-in amplifier or phase meter, and synchronized image frame grabbers, allow for higher spatial imaging resolution and recovery of the signal from the background through phase-sensitive techniques.

2.5.2. Laser-Based OBIC Photocurrent Calculation

The laser beam penetrating the silicon is electrically analogous to a forward-biased junction emitting minority carriers into the immediate

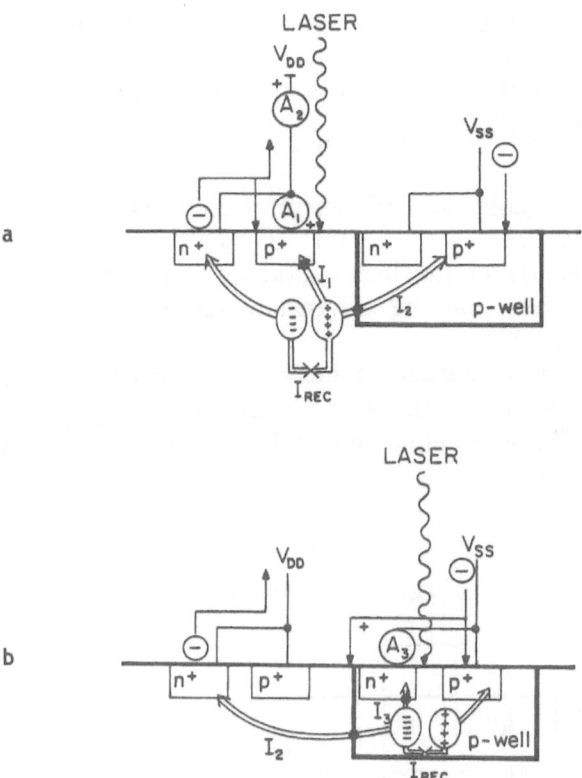

Figure 23. OBIC Photocurrents induced in CMOS structures by LSM: (a) outside of well; (b) inside of well.

vicinity of the beam (for He–Ne 6328-Å light, the absorption coefficient in silicon is $2\,\mu$m). For each light quantum absorbed in the silicon, a hole–electron pair will be produced. These photogenerated carriers will either be collected by surrounding junctions (giving rise to photocurrents) or recombine. Figure 23 shows the OBIC photocurrent collection when a laser beam hits outside a well (a) or inside a well (b).

The case of a p-well technology is considered here; all signs should be reversed for an n-well process.

There are only three possible photocurrents generated in the structure:

I_1: Photocurrent at the p–n–p "emitter" junction(s) (laser spot outside of well).

I_2: Photocurrent at the $p-n-p$ and $n-p-n$ "collector" junction. This is the well–substrate junction.

I_3: Photocurrent at the $n-p-n$ "emitter" junction(s) (laser spot inside of well).

The yield (or collection efficiency) of these photocurrents from the impinging laser beam will depend on such factors as process layer absorption, position of the beam spot, and minority carrier lifetime. These photocurrents can be electrically represented by the current sources in Figure 24.

The trigger level needed to induce latch-up can be readily found from Figure 24 with the criterion that the circuit loop gain must equal or exceed 1. The node equations at both collectors are (referring to

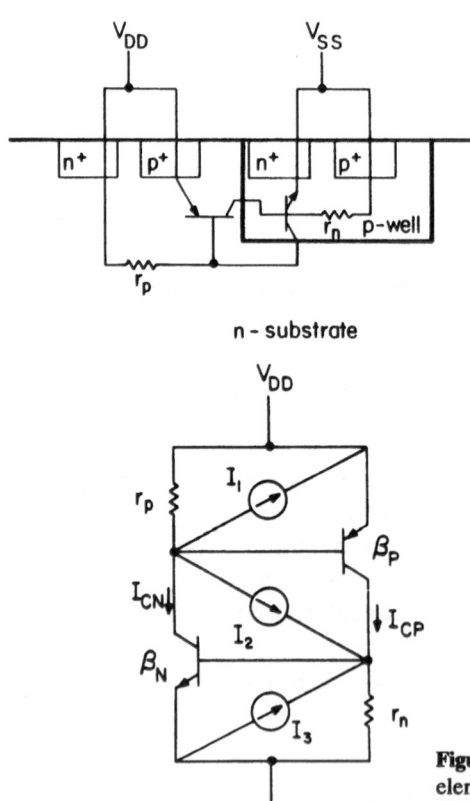

Figure 24. CMOS structure with parasitic elements due to laser-induced OBIC, and equivalent circuit with three possible laser-induced photocurrents.

Figure 24)[23] :

$$(V_{be}/r_p) + (I_{CP}/\beta_P) - I_{CN} = I_1 + I_2$$

$$(V_{be}/r_n) + (I_{CN}/\beta_N) - I_{CP} = I_2 + I_3$$

Solving for the parasitic transistor collector currents:

$$I_{CN} = (V_{be}/r_p) + (V_{be}/r_n \cdot \beta_p) + (I_{CN}/\beta_m \cdot \beta_p)$$
$$- \{I_1 + I_2[1 + (1/\beta_p)] + (I_3 + \beta_p)\}$$

$$I_{CP} = (V_{be}/r_n) + (V_{be}/r_p \cdot \beta_n) + (I_{CP}/\beta_n \cdot \beta_p)$$
$$- \{(I_1/\beta_n) + I_2[1 + (1/\beta_n)] + I_3\}$$

Setting $V_{be} = 0.6$ V (for latch-up to occur, both transistors must be turned on) and assuming $\beta_n, \beta_p \gg 1$ (worst case, small-geometry process), these equations become

$$I_{CN}(0.6/r_p) + (0.6/r_n \cdot \beta_p) - [I_1 + I_2 + (I_3/\beta_p)]$$

$$I_{CP} = (0.6/r_n) + (0.6/r_p \cdot \beta_n) - [(I_1/\beta_n) + I_2 + I_3]$$

These equations show the minimum transistor collector currents (I_{CN} and I_{CP}) needed to trigger latch-up. It can be shown that a sufficient condition for latch-up is met once *either* of the above current equations is satisfied.

2.5.3. Electron Beam-Induced (EBIC) Photocurrent Calculation (Figures 25 and 26)

It is possible, although complex, to derive an equation for the one-dimensional EBIC[24]; it explicitly includes the spatial dependence of the conduction/valence band and thus can be used for arbitrary doping profiles. The essential qualitative results are, however, that

- The EBIC decays exponentially, but only outside the space charge regions.
- The maximum of EBIC is not located at the crystallographic p–n junctions under certain conditions, but up to 10 μm away.
- The EBIC signal shapes depend on the electron beam current.

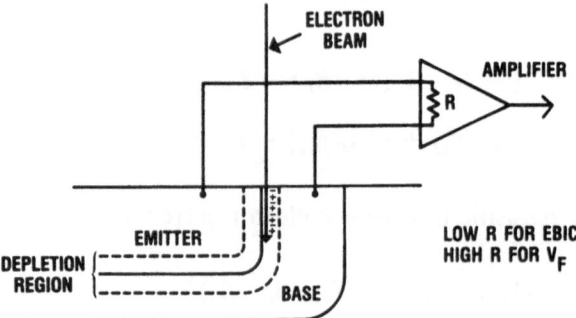

Figure 25. Cross section of junction, showing the generation of electron–hole pairs within the depletion region and the amplifier used either to collect the EBIC or to measure the forward junction potential.

2.5.4. OBIC and EBIC Image Processing

When a sequence of logical input stimuli are applied, two types of image processing take place, both preceded by thresholding (Algorithms Thresh-n), possibly filtering (Algorithms LUT-1 and Edge-1), and background subtraction (Algorithm Subtr-1).

2.5.4.1. Logic State Map[25,26]

For every conductive pattern, a logic value is measured for comparison with the expected value derived from CAD-based logic simula-

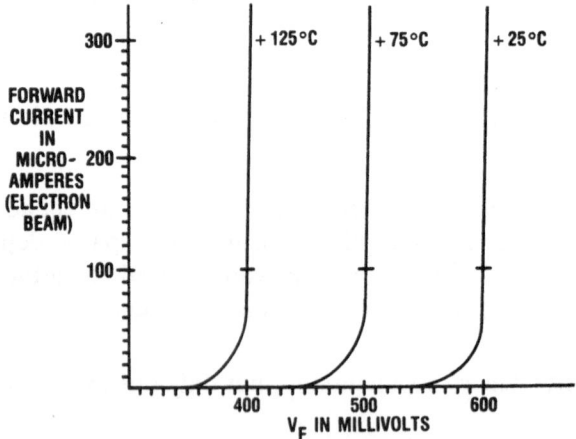

Figure 26. Characteristic curves showing the change in forward junction voltage with temperature. A primary electron beam current of 1 μA can generate an EBIC of from 100 to 200 μA, well beyond the "knee" of most junctions.

tion (prestored SPICE results, or equivalent). All top surface conductive patterns are displayed in different pseudo colors (Algorithm Pseudo-1) according to expected logical values. To prepare the logic state map, each interconnection pattern at the pattern level is linked to a circuit node on the logic-circuit diagram. A set of connecting lines linked to the same node is called a net. By using assigned net numbers, the inspection software correlates actual conductor patterns to nodes on the circuit diagram, and thus to simulated logic values.

Because of the number of input stimuli, test patterns, and imaging windows on the device under test, much processing is carried out in advance or off-line.

Moreover, the technique requires that the CAD design data bases allow for the circuit description at three levels, with correspondences between levels:

1. Circuit-to-circuit connections and types.
2. Placement and routing between circuits.
3. Pattern groupings per layer, with logic state attributes for each pattern.

2.5.4.2. Display of Photoinduced Current Image

Contrast in the photoinduced current image of a good device is compared with that of the device under test.[27] To pinpoint the defect, the image processing software (Algorithm Pseudo-1) displays the photoinduced current image of the good device, in green; the photoinduced current image of the device under test must appear in red. The areas that are functionally identical and thus good will appear yellow when the threshold images are superimposed. The malfunctioning patterns will appear in red or green, depending on the image contrast level.

Table 9. OBIC LSM Depth
Penetration in Silicon

Laser	Wavelength (nm)	Depth (μm)
He–Cd	325	0.01
N_2	337	0.01
He–Cd	442	0.2
He–Ne	633	2.0
GaAlAs	820	10.0
He–Ne	1152	>10,000

2.5.4.3. Adjustments

The beam penetration depths must be controlled (Table 9), especially when subsurface defects and logical states are to be accessed or avoided. In general, OBIC and EBIC analysis requires knowledge of the following:

1. *Devices:*
   - substrate material, insulating material
   - metallization material
   - line resolution and tolerance (minimum value)
   - voltage applied (range)
   - switching speed of circuit (range)
   - number of layers (range)
   - main defect types (size, geometry)
   - gate density (maximum value)
   - sensitivity of resist to optical or other illuminations
   - current density in metallizations (range in mA/mm^2)
   - current density underneath gates (range in mA/mm^2)
   - IC position tolerance during inspection (minimum value)
   - optical contrast between metallization and substrate at some given wavelength (minimum value)
2. *Electrical testing:*
   - current number of test stimuli per IC applied for full electrical testing
   - current electrical test time per IC (range)
   - test stimuli: voltage, frequency for all stimuli

Chapter 3

Metrology in Electronic Devices and Substrates

Metrology of critical dimensions (Figure 27) essentially involves measurement of linewidth, area, and occasionally height, down to the 0.01-μm range with positional repeatabilities of 0.03 μm or less.

3.1. LINEWIDTH MEASUREMENT

At present, straightforward linewidth measurement is difficult. Low-voltage in-line SEM[28] or confocal LSM[8] is required for accuracies of 0.25 μm or less. Fluorescence, confocal laser scanning,[29] microdensitometry, and digital processing of enhanced brightfield images are sufficient at for example, 1.5 μm. All approaches involve scanning a narrow optical slit across the magnified image of the features of interest. The light passing through this slit is measured by a photomultiplier tube (PMT) to form a microdensitometer profile (Figure 28).

In order to accumulate an image or several line scans (typically, 10–30) within an acceptable time, the object stage and/or the beam must be scanned at rather high frequencies. The generation of a 500 × 500 pixel image in 5 s involves 100-Hz scanning.

Thresholding or adaptive thresholding algorithms are then applied to the line scan PMT image to find the positions of the boundary points, of the boundary edges, or of clusters of boundary points.[30] Edge gradient correlation (Algorithms Edge-1 and Reg-1) is useful because of its insensitivity to reflectivity changes caused by layer thickness variations, especially with Gaussian filtering kernels.

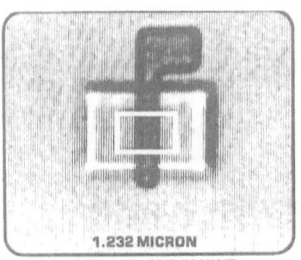
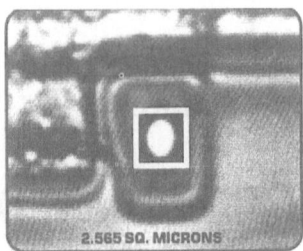
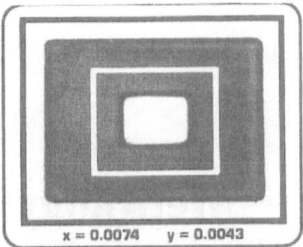

LINEWIDTH MEASUREMENT
This feature on a multi-layer wafer was
measured to be 1.232 μm.

AREA MEASUREMENT
Contact area measurements such as this one
of 2.565 μm² can be utilized as part of
pass/fail inspection strategy.

MISREGISTRATION MEASUREMENT
X and Y misregistration values provide key
information necessary for controlling the
alignment process.

Figure 27.   Examples of metrology in wafers. (Courtesy KLA Instruments Corporation.)

A measure of the distance (maximum or minimum) between such points or edges gives the linewidth at the slit location or within the selected area of interest.

A sequence of image-processing algorithms for linewidth measurement is presented in Table 10.

The linewidth accuracy is highly dependent on the illumination, and

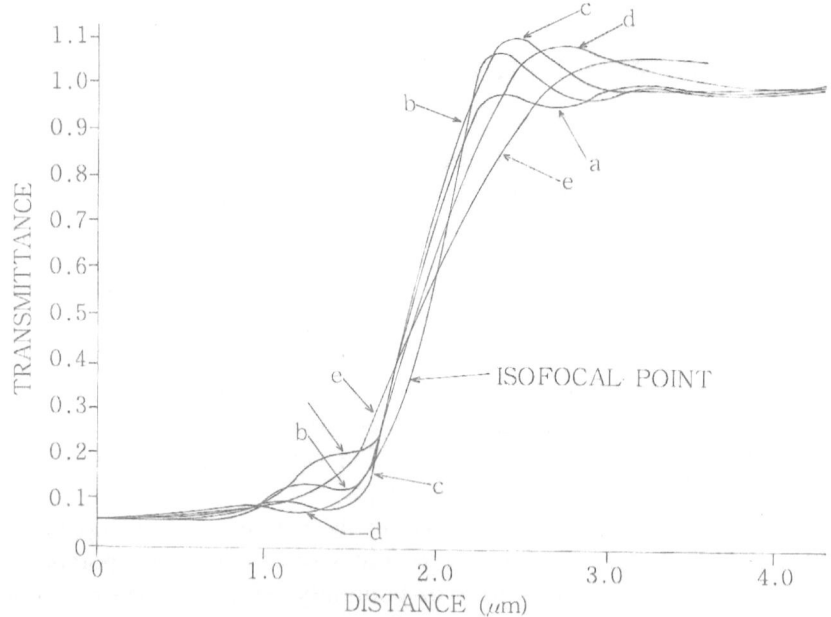

Figure 28.   Variance of light transmittance across an edge for metrology, and dependence of focusing; the isofocal point occurs at about 35% and is least subject to focusing changes.

Table 10. Basic Steps in Linewidth Measurement

Sensor: Video camera, SEM, LSM, or confocal laser.
Attitude: Top-down view.
Illumination: Collimated light, laser, or *e*-beam.

0.1 Optical slit or narrow window illumination.
0.2 Retrieve nominal line edge positions and width values, and register with the sensor position.
0.3 Geom-2: calibration of physical pixel dimension.
1.0 LUT-1: selection of LUT for physical edge enhancement.
2.0 Thresh-1: thresholding of image in step 1.
3.0 Edge-*n*: edge detection in step 2, using step 0.2 as initial information.
4.0 Offset-1: offset or edge-to-edge distance determination.
5.0 Translate the distance from step 5 into a physical dimension using step 0.3.

---

fluorescent light (optical, laser, brightfield) has the following advantages:

1. Optical lens chromatic aberration is less because fluorescence occurs over a narrow wavelength range.
2. Measurement precision is improved due to the high contrast images. A fluorescent line emits light against a black background.
3. Measurement errors caused by sublayer and substrate reflections are eliminated.
4. Since the emitted fluorescent light is essentially incoherent, no diffraction fringes appear in the image profile. This allows high-accuracy measurements to be made down to the diffraction limit of the optical system.
5. Lastly, dark lines, which appear in an image created by monochromatic light at film thicknesses that are multiples of the illuminating wavelength, are absent with fluorescent measurements.

Fluorescent linewidth images do, however, have low light levels which make it difficult for normal television cameras to generate useful measurement profiles. However, PMT-based systems have been shown to have sufficient sensitivity for this application.

In general, all metrology systems must have computer control of illumination (intensity, slit location) and autofocus as an option.

Some systems, more manual in essence, operate by "just-touch" settings between superimposed images (Figures 29 and 30).

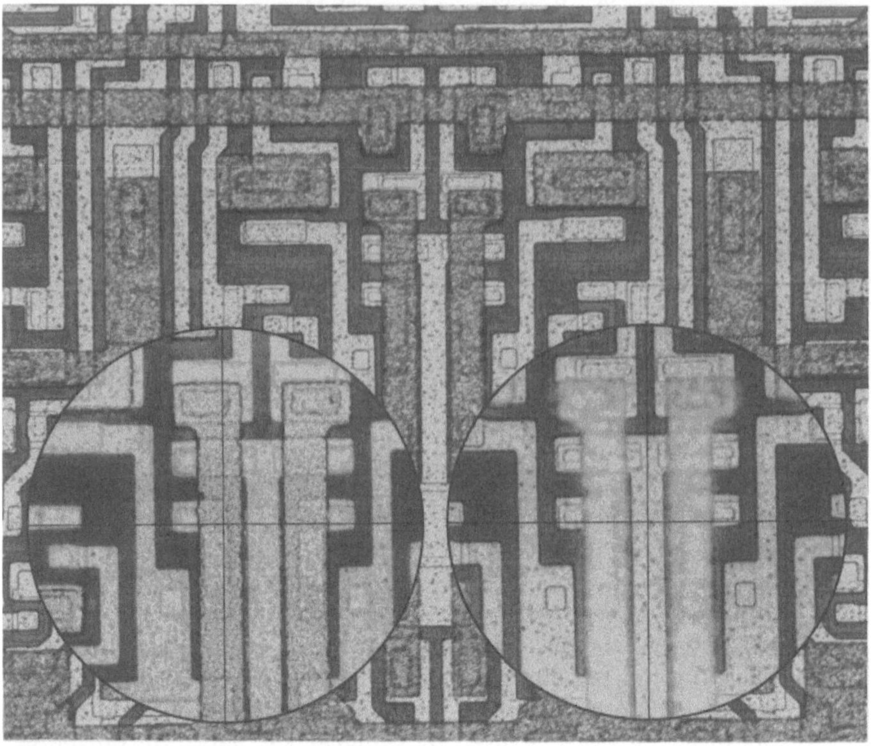

Figure 29. Alignment and focusing for linewidth measurement by optical microscope. (Courtesy Leitz.)

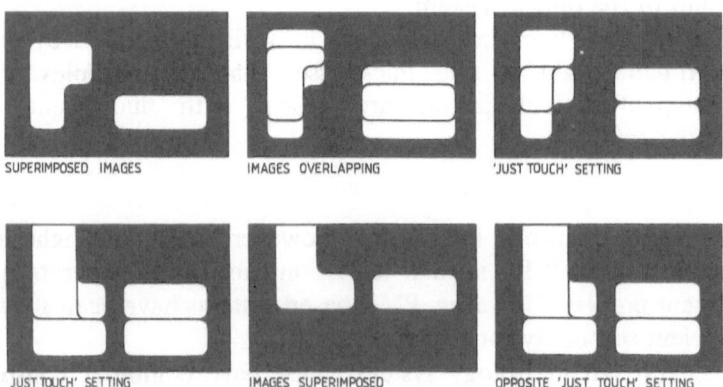

SUPERIMPOSED IMAGES   IMAGES OVERLAPPING   'JUST TOUCH' SETTING

JUST TOUCH' SETTING   IMAGES SUPERIMPOSED   OPPOSITE 'JUST TOUCH' SETTING

Figure 30. "Just-touch" manual settings for alignment, linewidth, and surface measurements; image overlay and the image processor display are employed.

3.2. AREA MEASUREMENT

Area measurement essentially has image segmentation (Chapter 17) as a prerequisite, either by fixing an area of interest (AOI) independent of actual image features or by true image segmentation (regions or connected components). In the image segment, the area can be determined by either thresholding (Algorithm Thresh-n), followed by pixel counting of all pixels above or under the threshold, with finally scaling by the physical pixel size, or by binary morphomathematical processing, in order to determine the pixel count by erosion operations (see Chapter 18).

3.3. SURFACE FLATNESS AND PROFILING

3.3.1. Flatness Measures

Surface flatness is measured by either of the three following criteria applicable to the elevation $h(i, j)$ on the surface:

1. *Total indicated reading (TIR):* TIR is the difference in elevation between the highest and lowest points on the surface, expressed as the absolute value.
2. *Nonlinear thickness variation (NTV):* Given a neighborhood window W around the center of the surface $(0, 0)$, NTV is the maximum of the height deviations between $(0, 0)$ and the median height of diametrically opposed points with respect to the center in the window:

$$\text{NTV} = \underset{i,j}{\text{Max}} \, |h(0, 0) - [h(i, j) + h(-i, -j)]/2|$$

3. *Focal plane deviation (FPD):* FPD is the peak absolute deviation of the surface from a reference plane (established by points on the wafer periphery).

See also Section 6.2 for other flatness measures.

3.3.2. Flatness Measurement

A variety of methods for flatness measurement exist, such as grazing angle laser interferometry, ultrasonic probing, capacitance probing, and laser scan with chuck rotation.[31]

3.3.3. Flatness Display Images

The flatness tester will display the raw elevation image $h(i, j)$ for all points (i, j) on the surface, eventually with sampling and pseudo-color coding of the raw elevation values h (Algorithm Pseudo-1).

In overlap to such an image, the following features are displayed:

1. Peak local slope image, which is the smoothed gradient image of the elevation image $h(i, j)$ (Algorithm Edge-1); the gradient or slope intensity can itself be pseudo-color coded (Algorithm Pseudo-1).
2. Contour mapping, which links together by curves pixels in $h(i, j)$ of equal elevation range values; this may use Algorithm Bound-2, applied directly to the h-image.
3. Isometric displays, which are three-dimensional views to show the shape of the surface, with deviation from planarity displayed to a

Figure 31. Flatness vision test result on a SMD package. (Courtesy Applied Intelligent Systems.)

desired degree of magnification; this requires display technology or computer graphics.

3.3.4. Areas of Application

Flatness measurement is essential for semiconductor wafers (see Chapter 6), as well as for multilayer printed wired boards (additive or subtractive plating techniques) (see Section 10.9) and for SMD packages (Figure 31).

Chapter 4

Inspection of Integrated Circuits and Gate Arrays

The importance of visual inspection of ICs, usually carried out by skilled human operators, stems from the fact that most electrical or mechanical device defects can be revealed by optical features visible in, for example, a microscope.

Key to optical inspection of ICs is that the task is dominated by geometrical patterns and that defect or geometry feature size is very small, while the total die area to inspect is large, thus leading to the requirement to inspect manually or automatically very large fields of view (Table 11).

A list of IC inspection tasks is presented in Table 12.

4.1. INSPECTION STANDARDS

A number of standard inspection procedures exist. Examples are MIL-STD-883-D and its Method 2009-1 (external visual), Method 2010-3 (monolithic internal visual), Method 2017-1 (internal visual hybrid), and Method 5004, as well as MIL-STD-1772 (Certification Requirements for Hybrid Microcircuit Facilities)[32] and ESA, NATO, or national equivalents. Each method specifies a series of visual acceptance criteria for a range of defect types (Figure 32). The order in which such verifications must be carried out is not specified.

Today, compliance with the test standards is based on manual procedures. Clients may provide waivers on a project-by-project basis to

Table 11. Number of Fields of View Required to Inspect a Die for a Given Feature Size and Area[a]

Die area (sq. mils)	Feature size (μm)	Objective	Fields of view
10,000	5.0	40×	30
62,500	2.0	100×	1200
90,000	2.0	100×	1700
90,000	1.25	160×	4400
100,000	1.00	160×	7500

[a] Assumptions: 18-mm field; 10× eyepiece; visual resolution: 2′ of arc; required resolution element = feature size/12.

Table 12. IC Inspection Tasks[a]

Type of inspection	Inspection tasks
Metrology and geometrical sorting	Linewidth measurement Film thickness Surface flatness Pattern alignment Surface profiling
Surface inspection	Conductor comparison Detecting particulates Pattern anomalies Internal/external visual
Process control	Doping concentration Resistivity Step coverage Chemical vapor analysis Surface specularity
Defect analysis	Impurity analysis Film composition 3-D viewing

[a] Adapted from VLSI Research.

allow for automated inspection instead, where it would be advantageous. It is likely that in the future, the 100% human inspection requirement will be relaxed, thus allowing for many more automated tasks.

4.2. INSPECTION PROCEDURE IMPLEMENTATION

The inspection procedures (e.g., MIL-STD-883, Method 5004; MIL-M-38510; MIL-S-19500; MIL-STD-750) involve sampling plans,

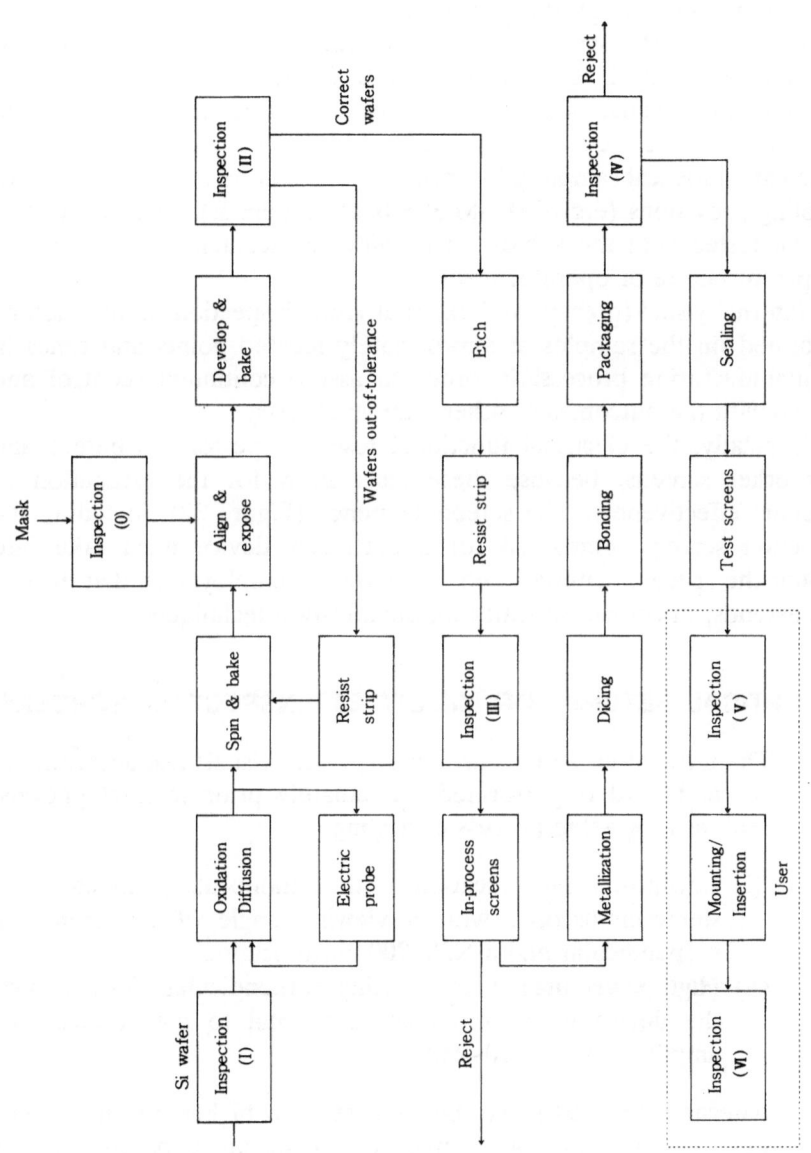

Figure 32. Sequence of inspection and other fabrication steps throughout IC manufacturing.

with the selection of inspection sublots which consist of circuits or boards of a single type covered by a single detail specification, with manufacturing on the same production line; this requirement extends all the way to final seal by the same production technique, to the same design rules and package, and to the same material requirements. The device class must also be given: Class B, Class C, Class S, or JAN.

First, all circuits, boards, or devices should be marked by a code indicating the date the lot was submitted for screening and inspection.

Next, statistical sampling for inspection should be in accordance with sampling provisions (e.g., MIL-M-38510, Appendix B); reserve samples may be tested with the sublots to provide replacements in the case of equipment failure or operator error.

Internal visual (precap) and external visual inspection should then be performed on the samples at appropriately located points and times in the manufacturing process, in order to assure continuous control and overlap with the outcomes of other mandated tests.

Typically, the electrical functional tests are performed before and after other screens, because these tests allow for the estimation of screening effectiveness. The screen sequence (Figure 32), including the inspection screens, employed for a particular device must take into account the specific materials and configuration employed in that device. This extends, of course, to wafer lot qualification techniques.

4.3. OPTICAL SETUPS FOR MIL-STD-883 INSPECTION SCREENS

1. The internal visual die (precap or preseal visual) test according to Method 2010 is performed immediately prior to final package seal, as a two-step process consisting of

   (i) Low-power magnification, with a monocular, binocular, or stereomicroscope, with a viewing angle of 30° from the perpendicular and a ×30–200 magnification.
   (ii) High-power magnification using perpendicular viewing, with the device under illumination normal to the surface; the amplification is ×70–280.

   Alternatively, SEM may be used at even higher magnifications, but only as part of sample lot qualification because of its destructive nature (Table 13).
2. The external visual inspection (postcap) test according to Method 2009 essentially involves vertical or lateral low-level magnification, sufficient to check the package and leads.

Table 13. Optical Microscopy versus SEM Microscopy for Wafer Processing[a]

	Minimum linewidth (μm)		Sensor	
Device type	Production	R & D	Production	R & D
LSI (silicon)	0.9	0.7	Optical	Optical, Scanning ion beam
Bipolar ICs (silicon)	1	0.5	Optical	SEM
Transistors	0.8	0.5	SEM	SEM
GaAs ICs	0.8	0.3	SEM	SEM, LSM
GaAs FETs	0.3	0.25	SEM	SEM

[a] Ref. 33.

3. The X-ray radiographic inspection according to Method 2012 examines packages for debris and screens die attach eutectics. The specific rejection criteria are interpretation dependent, and metal packaging blocks the X-rays from less dense contaminants. Gold wires are, on the other hand, well detected.

4.4. TEST PATTERNS

In all cases, test patterns may be used to provide convenient areas for probing (Figure 33). These areas may be included within the actual circuit die area to ensure the necessary matching of characteristics. In addition to providing locations for probing with minimum damage to the actual circuits, the test pattern may be designed to provide components which amplify the characteristics to be measured.

4.5. OPTICAL DEFECT FEATURES

In terms of compliance with a proper reference circuit, the optical features to be checked include (Figure 34)

- Offset of three-dimensional structures.
- Calculated prolongation areas or areas between polygonal structures.
- Scratches in a structure and their distance to its boundaries.
- Scratches revealing oxide and area exposed.
- Location, surface, and shape of pad narrowings.

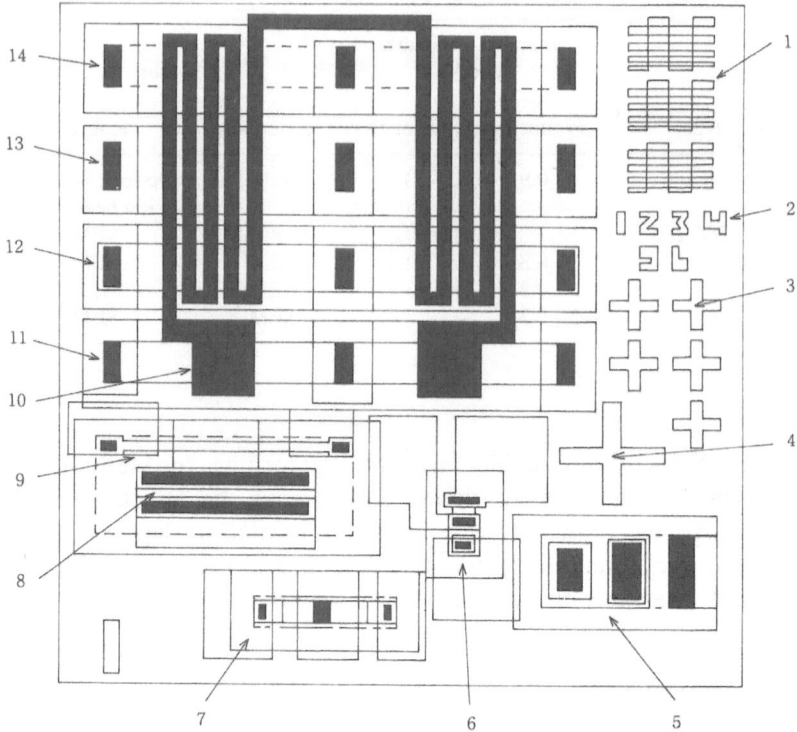

Figure 33. Test pattern: (1) resolution patterns for six masks; (2) mask identification; (3) progressive alignment key (for example, ISO to BL, base to ISO); (4) master alignment key (all masks aligned with respect to each other); (5) large probe transistor; (6) typical small-signal transistor; (7) squeezed resistor (base width); (8) base diffused resistor, $l/w < 1$; (9) base diffused resistor, $l/w > 1$; (10) metallization resistance; (11) sheet resistance of emitter; (12) sheet resistance of base; (13) sheet resistance of epitaxial deposition; (14) sheet resistance of buried layer.

- Misalignment between structures.
- Diffusion bands between zones.
- Overspill of pads toward the outside.
- Welding spot shape anomalies.
- Fillets and crowns around structures.

These optical features in turn relate to defects and to manufacturing processes (Table 14). About 120 defects are specified in the standards, grouped into the following classes:

1. Metallization: scratches, voids, corrosion, adherence, bridges, alignment.

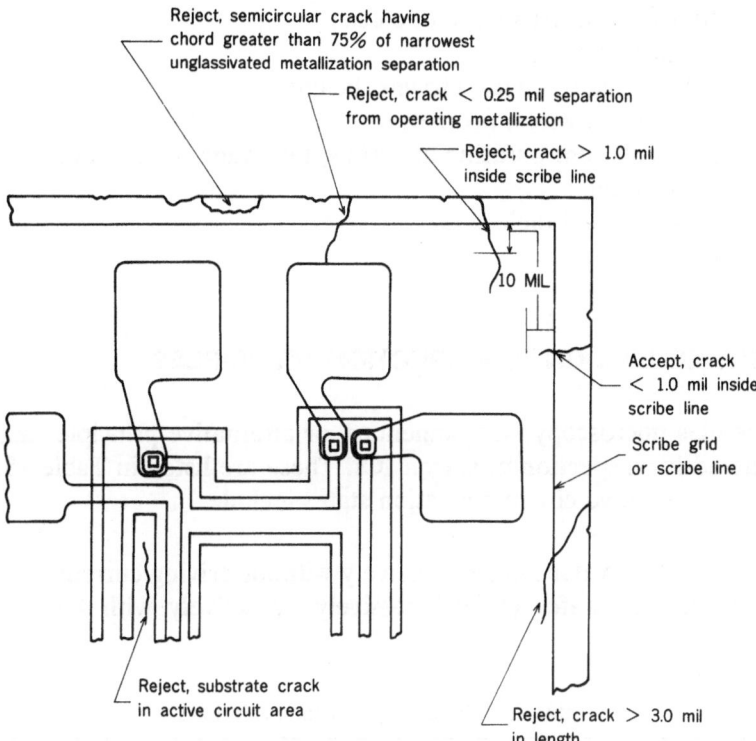

Reject, semicircular crack having chord greater than 75% of narrowest unglassivated metallization separation

Reject, crack < 0.25 mil separation from operating metallization

Reject, crack > 1.0 mil inside scribe line

10 MIL

Accept, crack < 1.0 mil inside scribe line

Scribe grid or scribe line

Reject, substrate crack in active circuit area

Reject, crack > 3.0 mil in length

Figure 34. Visual inspection criteria examples, as defined in MIL-STD-883, Figure 2010-15.

Table 14. Typical IC Defects by Circuit and Process Type

Process type	Defects
CMOS	Junction defects (contamination, leakage, short circuit) Missing implant Lithographic defects
Bipolar	Open wire bonds Metallization microcracks at oxide steps Electromigration
Linear	Mobile Na^+ charges Surface or interfacial charge from biased metal conductors Open wire bonds Metallization microcracks at oxide steps

   2. Diffusion and passivation defects.
   3. Substrate defects.
   4. Welding: gold, edges, crowns, beams.
   5. Conductors: wires, geometry.
   6. Encapsulation: parasitic materials, mounting, orientation.
   7. Vitrification.
   8. Insulators and dielectrics.
   9. Resistive films.

4.6. OTHER SILICON IC INSPECTION PRINCIPLES

Besides microscopy, supplementary or alternative principles may be considered for inspection of silicon ICs. These are listed in Table 15.
The alternative circuit activation states include

- Circuit or wafer imaged passively without driving current.
- Circuit or wafer imaged passively or actively with DC driving current.

Table 15.   Silicon IC Inspection Principles

Imaging principle	Use for silicon IC inspection[a]	Reference(s)
X-ray microradiography (XMR)	++	
Laser scanning imagery (LSM)	+++	
Optical laser beam-induced currents (OBIC)	++	
3-D structured lighting	+++	
Scanning electron beam imagery (SEM)	+++	
Scanning transmission electron imagery (STEM)	+++	
Electron beam-induced currents (EBIC)	++	
Stroboscopic contrast SEM	++	
Scanning ion beam imagery	+++	
Scanning Auger microscopy (SAM)	++	34
High-repetition laser pulse imagery	+	
Scanning tunneling microscopy	++	
Interferometry and speckle	+	
Reflectance maps	++	
Hot spot detection by liquid crystals	+	35
Ellipsometry	+	
Raman imaging spectroscopy	++	36, 2
Secondary ion mass spectrometry (SIMS)	+	
Fourier transform photoluminescence	+	2

[a] The usefulness for silicon IC inspection is graded from + (lowest) to +++ (highest).

- Circuit or wafer imaged passively or actively with AC driving circuit, and through a sequence of logical states programmed from a stimuli/test word generator.

For example, laser scan imagery (LSM) (see Sections 2.3 and 2.5) or scanning ion beam imagery can reveal logic states, and their reversible changes, through optical beam-induced currents (OBIC); structures with nonlinear behavior, e.g., inverters; photoresponse of the structures, including laser-induced fluorescence (LIF); topography of the superficial structures, with penetration depth of about $3 \mu m$ in silicium at 633-nm illumination wavelength, $10-30 \mu m$ at 1150-nm illumination wavelength, and $10-30 \mu m$ at 1150-nm illumination wavelength; and most geometrical features listed in Section 4.3.

In turn, scanning electron microscopy (SEM) (see Sections 2.2 and 2.5) offers positional accuracies of 1%, repeatability of 1%, and 100–250-Å resolution for 3–30-mG magnetic fields and 0.2–2-keV electron beam acceleration voltages; the magnification range is continuous from ×10 to ×300,000, and the depth of field very high, thus allowing for the imaging of a complete structure at one time. Sample preparation problems limit SEM for cross-sectional studies to a maximum magnification of ×100,000.

X-ray microradiography (XMR) can now magnify images by up to about ×50–×200 in real time, giving excellent surface and subsurface features.

Table 16 gives a further comparison of these three types of scanning microscopy.

4.7. OTHER III–V COMPOUND IC INSPECTION PRINCIPLES

Besides microscopy, supplementary or alternative principles may be considered for GaAs, InP, and similar III–V compound ICs[37] as well as related photodetectors.[38] These are listed in Table 17.

4.8. RESULTS OF IC INSPECTION AND LINK TO OTHER TEST OR DEFECT ANALYSIS METHODS

Visual inspection of ICs by manual or automated means must be supplemented by other analysis or quality control methods, such as

- Statistical lot acceptance sampling plans.
- Stabilization bake.

Table 16. Comparison of the Three Typical Scanning Microscopies

	Scanning ultramicroscopy	Scanning electron microscopy (SEM)	Scanning X-ray microscopy
Source of monochromatic beam	Laser	Electron gun	Characteristic X-ray
Image formation	Good optical lenses	Electron optical lenses	None
Mode of image formation			
(a) Elastic scattering	Rayleigh scattering	Backscattered electrons	Bragg and Laue diffraction (Lang topography) Anomalous dispersion
(b) Inelastic scattering	Raman scattering, Brillouin scattering	Auger electrons, secondary electrons	
(c) Luminescence	Photoluminescence	Characteristic X-ray, cathode luminescence	Fluorescent X-ray
(d) Induced current	Photoelectric current	Electron beam-induced current (EBIC)	
(e) Absorption	Microspectroscopy	Absorption current	OK
(f) Transmission	OK	STEM	OK
Objects	Inhomogeneity of refractive index	Surface structures	Strain gradient in crystals
Thickness of sample	From 0.5 mm to a few cm	About 1 mm or less	Less than 1 mm; dependent upon the absorption coefficient
Resolution	A few μm	A few nm	A few μm
Merits	(a) Nondestructive investigation	(a) High resolving power	(a) Fitness for characterization of defects in crystals
	(b) No introduction of defects by handling and processing	(b) (Semi-)quantitative analysis of elements on surfaces	
Weak points	Transparent specimen only	Clean surfaces only and/or extremely thin specimens	Crystals only

Table 17.   III–V Compound Inspection Principles

Imaging principle	Use for III–V IC inspection[a]	Reference
IR laser scattering	++	
Laser photoluminescence	++	2
Scanning electron beam imagery (SEM)	++	
Synchrotron radiation topographic imaging	+	
X-ray microradiography (XMR)	+	
Optical laser beam-induced currents (OBIC)	++	
Electron beam-induced currents (EBIC)	+	
Photoinduced microwave reflection	+	
Cathodoluminescence from SEM	( )	
4.2 K photoluminescence	( )	39
Differential scanning calorimetry	( )	

[a] The usefulness for III–IV IC inspection is graded from ( ) (unknown) to + (low) to +++ (highest).

- Thermal shock, mechanical shock, acceleration, and temperature cycling.
- Reverse bias.
- Hermeticity tests on encapsulation.
- Electron spectroscopy (ESCA).
- Photovoltage spectroscopy (PVS).
- Deep-level transient spectroscopy (DLTS).
- X-rays.
- Energy dispersive X-ray spectrometry (EDX) (1-μm resolution).
- Wavelength dispersive X-ray spectrometry (WDX).
- X-ray fluorescence analysis (XRF).
- Atomic emission spectroscopy.
- Surface resistivity probes.
- Robinson backscattered electron imaging (RBEI) (20-nm resolution).
- Rutherford backscattering spectrometry (RBS).
- Laser microprobe mass analysis, with an ionizing pulsed laser, and time-of-flight mass spectrometry.
- Differential deep-level transient spectrometers.

4.9. SURFACE AND DEPTH ANALYSIS OF SEMICONDUCTORS

Some of the techniques which may be used for surface and depth analysis of semiconductors are described in the following sections. A

Table 18. Comparison of Surface Analysis Techniques[a,b]

Parameter	Electron probe			Photon probe: ESCA[f]	Ion probe: SIMS[g]
	RBEI[c]	EDS[d]	SAM[e]		
Surface depth analyzed	1 μm	1 μm	1 nm	1 nm	3 nm
Information provided	No profiling Compositional variations	No profiling Elements	Profiling Elements	Profiling Chemical bonding	Profiling Isotopes
Spatial resolution	20 nm	1 μm	1 μm	1 mm	20 μm
Quantitative accuracy[h]	None	2%	10%	2%	25%
Sensitivity	0.5%	0.1%	0.1%	0.1%	0.01–0.0001%
Relative costs	1	1	2	2	4
Exceptions	No elemental identification	No light elements below sodium	No insulators	Low spatial resolution	Spectral interferences, high cost

[a] Ref. 40.
[b] For all these techniques, the information obtained depends on the specimen, variations among instruments, and the combination of information required.
[c] RBEI: Robinson backscattered electron imaging.
[d] EDS: Energy dispersive X-ray spectrometry.
[e] SAM: Scanning Auger microanalysis.
[f] ESCA: Electron spectroscopy for chemimal analysis.
[g] SIMS: Secondary ion mass spectrometry.
[h] Referenced against known standards.

Table 19. Basic Research on IC Microanalysis and Surface Analysis with Imaging

Technique	Remarks
Scanning ion microscopy (SIM)	50–100-μm ion beam spot size; large depth of field; high contrast; resolution of less than 0.1 μm
High-resolution scanning secondary-ion mass spectrometry	Resolution of 0.1 μm in secondary electron and ion images
Stroboscopic voltage contrast SEM	Problems of voltage resolution and interpretation
Auger microscopy	Parallel 1-s 1024 × 1024 detection scheme
Scanning electron acoustic microscopy (SEAM)	Subsurface imaging at 0.7-μm resolution (at best)

comparison of five surface analysis techniques is presented in Table 18 and research techniques employed can be found in Table 19.

4.9.1. Wavelength Dispersive X-Ray Spectrometry (WDX)

Widely used, the WDX technique operates on the principle that when a beam of X-rays is directed onto the surface of a specimen, secondary (or fluorescent) radiation is emitted by the specimen, containing wavelengths which are qualitatively and quantitatively characteristic of each element present.

In a sequential spectrometer, this secondary radiation is converted by a collimator into a parallel beam and directed onto an analyzing crystal in order to separate out the wavelengths. These, in turn, are reflected into a scanning radiation detector mounted on a precision goniometer. At certain angles which correspond to each wavelength present, a peak of radiation intensity is obtained. Just as the angle at which a peak occurs is uniquely related to a particular element, the intensity of the peak is proportional to the concentration of that element in the sample.

In a simultaneous spectrometer, individual element channels, each with its own crystal and detector, are mounted in concentric circles around the sample, enabling all elements to be measured at the same time.

4.9.2. Energy Dispersive X-Ray Spectrometry (EDX)

Energy dispersive X-ray spectrometry is increasingly used for non-destructive testing. An excitation source, which can be X-rays or the beam of an electron microscope, is used to bombard the specimen. The resulting interaction of the beam with the specimen can be accompanied by the release of an X-ray photon, the energy of which is characteristic of the element from which it originated. Individual photons are collected by a detector made of lithium-drifted silicon, which can be built into an X-ray spectrometer or an electron microscope column. This converts the energies of the photons into electrical pulses, measurement of which provides a means of identifying the specimen's constituent elements. A photon count enables the relative quantities of each element to be established.

4.9.3. Optical Beam-Induced Currents (OBIC)

A recently launched alternative to the well-established EBIC (electron beam-induced current) technique, the OBIC technique (see Section 2.5) is available on some laser scanning microscopes (LSM). The technique creates electron-hole pairs by photon excitation near the rectifying junction of a semiconductor. The current locally induced from the creation of charge carriers by the scanning laser spot is collected and monitored. This current is then used to modulate the brightness of a scanning spot on a cathode-ray tube to give a "map" of electrical activity.

4.9.4. Photovoltage Spectroscopy (PVS)

In PVS, the photovoltaic response of a semiconductor material is measured as a function of the wavelength of the light illuminating the material. The spectrum obtained can be used to determine the bandgap of the semiconductor.

In compound semiconductors, the bandgap gives a measure of chemical composition. The most important application of this is the measurement of aluminum contact in AlGaAs devices such as laser diodes, high electron mobility transistor (HEMT) structures, and many optoelectronic systems.

4.9.5. Deep-Level Transient Spectroscopy (DLTS)

DLTS measures temperature-dependent transient capacitance signals from deep states which are either optically or electrically filled and then allowed to empty.

These deep states are either intentionally added or are the result of problems associated with the production of a particular device. Intentional deep states increase the resistivity of materials and substrates such as GaAs.

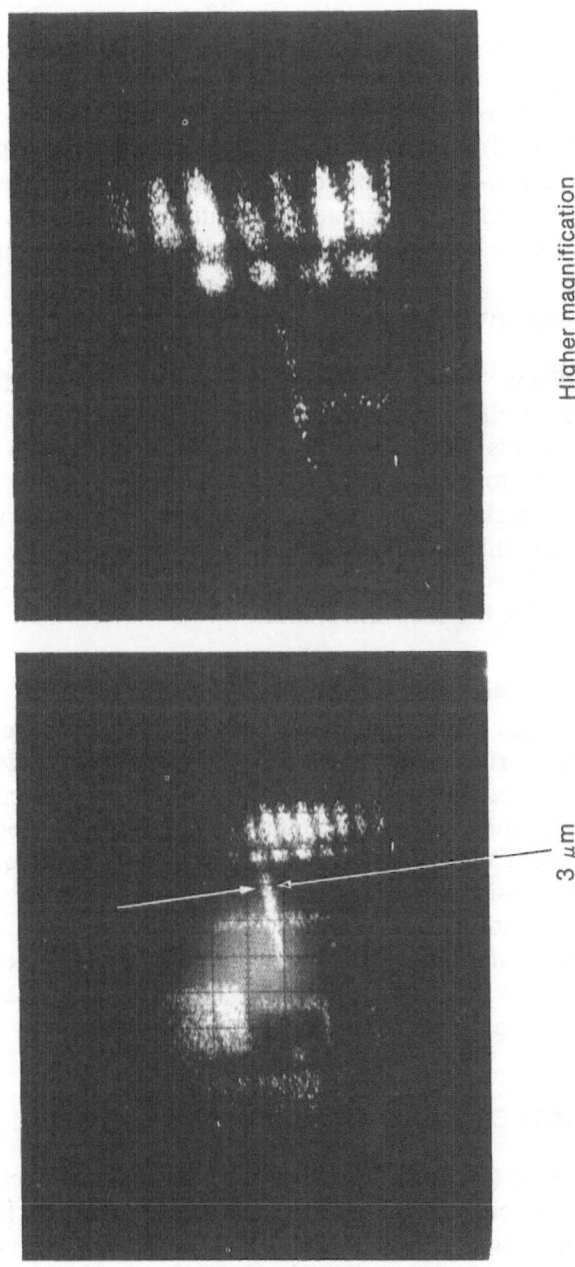

Higher magnification

3 μm

Figure 35. SIMS images, for doping analysis; ^{11}B was implanted on a Si wafer through a mask, to achieve an R_p concentration of 10^{18} cm^{-3}; analysis conditions are: O_2^+, 13 keV, 5 nA. (Courtesy Riber S.A.)

4.9.6. Secondary Ion Mass Spectrometry (SIMS)

Secondary ion mass spectrometry provides parts per billion sensitivity and precise spatial information (Figure 35). A beam of microfocused ions is used to knock off, from a small point on the sample, successive atomic layers which are subsequently analyzed by a high-performance mass spectrometer. Scanning the ion beam over a small area enables a series of two-dimensional chemical maps to be built up over increasing depth. Stored in a computer, this information can be manipulated to give color images representing detailed chemical distributions.

4.9.7. "Time-of-Flight" Mass Spectrometry

In "time-of-flight" mass spectrometry, a small cloud of ionized sample is electrostatically accelerated down a flight tube, where, because heavy and light atoms travel at different speeds, the atoms become separated according to mass. Thus, by measuring intensity and the respective time of flight, a detailed mass spectrum is built up. This method of analysis offers advantages of high mass range, uniform transmission for all ions, parallel detection of all masses, and a detection sensitivity approaching the theoretical limit.

4.9.8. Scanning Auger Electron Spectroscopy (SAM)

In scanning Auger microanalysis, an energetic beam of electrons is fired at the sample surface and, following a particular rearrangement of the electronic levels of the component atoms, secondary electrons are ejected and energy analyzed. On account of the very low penetration depth of the incident electrons (2–10 nm), the Auger technique is an extremely surface sensitive technique which provides detailed quantitative elemental information.

Focusing and rastering of the beam across the sample surface produces a secondary electron image. The beam is usable also as a scanning microprobe in an elemental mapping mode.

4.10. BUBBLE MEMORY INSPECTION

Bubble memories (Figure 36) have data blocks that are electronically addressable without consideration for interblock spaces or for block headers. Also, they can be turned on from a powered-off condition in about 10 μs to retrieve or store a single byte. The bubbles, resting on strong permalloy poles, are moved around by the rotating magnetic fields (Figure 37). Error rates depend on how close the operating field is to

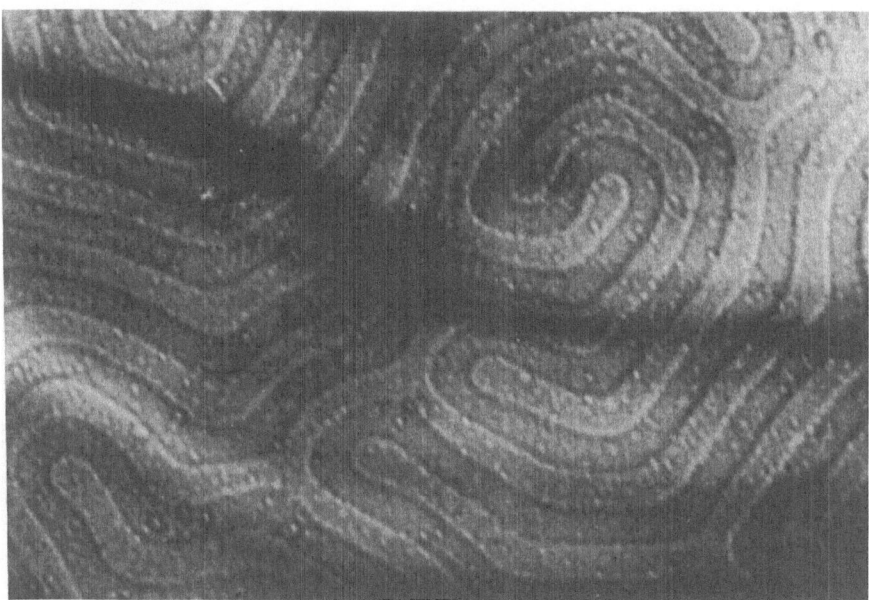

Figure 36. STEM image of the magnetic domain structure of a bubble memory (thinned samarium-thulium-iron-oxide-garnet crystal); the resolution is 0.5 μm, and reveals the wall structure, invisible by optical microscopy.[237]

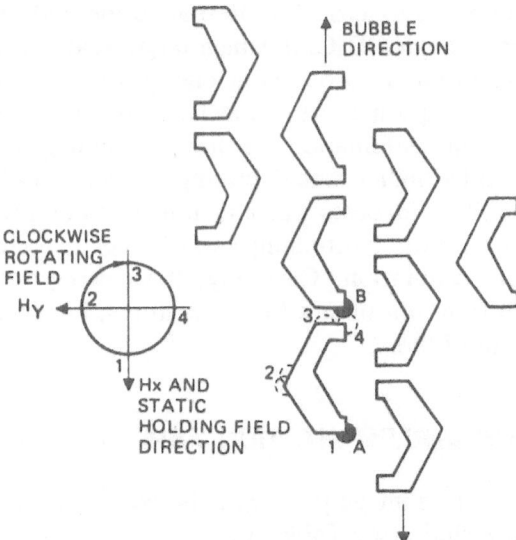

Figure 37. Propagation pattern in a bubble memory; as the magnetic field is rotated clockwise one full revolution, the bubble at location A will move one step to location B.

bubble collapse or to bubble strip-out, when the bubble domain stretches between permalloy elements.

All these operating conditions point directly to dynamic fault imaging (DFI) (Section 2.2.4) for quality control, usually with single- or dual-wavelength laser scanning microscopes (LSM) (Section 2.3).

Another approach consists in applying a 0.1-μs pulse magnetic field (of reverse polarity to the saturating magnetic field), followed by polarized laser light illumination. The unstable local areas in the layer containing bubbles are revealed by the reflected laser light. Such defects include local strains, heterogeneities of permalloy elements, gaps between elements, and alterations of the local anisotropy field.

4.11. LASER TRIMMING AND LINK CUTTING

Trimming involves the continuous adjustment of thin film resistors for functional adjustments. Link cutting corrects fault-tolerant memory circuits. Mask repair consists in reworking mask defects. Many analog as well as digital IC manufacturers today use these techniques for yield improvement.

These techniques involve cutting, by one or several laser pulses, a number of conductive links that are 1.5–2.0 μm wide. Alignments are made by measuring changes in the reflected laser energy as the laser is scanned across feature surfaces. An alternative method uses photoconductive alignment through OBIC, in which target features are located by measuring changes in the absorbed laser energy via p–n junctions in the device under test. The die sizes can be as large as 1.1 $\times$ 0.5 cm.

The demand for automation of laser trimming through image processing is driven by higher manufacturing volumes and by the wish to reduce setup time. This involves wafer identification and tracking, wafer and die alignment, beam positioning control, beam focus and energy control, and beam calibration. Currently, the existing laser cutting or mask repair systems do not offer all these features, but image processing is a necessity for all of them.

4.12. INSPECTION IMPLEMENTATION ASPECTS

The main uses of an image processing-based IC, ASIC, or gate array inspection system include (see Table 20):

1. Acceptance quality control at end user sites.
2. Production quality control at manufacturer sites.

Table 20. Basic Steps in IC Inspection Algorithm

Sensor: See Sections 4.6 and 4.7.
Attitude: Top-down view of IC.
Illumination: See Sections 4.6 and 4.7.

0. Geom-1.
1. Align-1: alignment with reference CAD layout.
2. Align-1: fine alignment with anchor points.
3. Thresh-1 (subtraction from reference, and thresholding).
4. Edge-1 or Filter-1.
5. Pseudo-1: false color or pseudo 3-D representation of step 4.
6. Segm-1 with feature extraction.
7. Metrology of features in step 6, with comparison to reference or CAD values.
8. Graphic highlighting of defect locations and types.
9. Knowledge-based expert system processing of defects in step 7 with reference to manufacturing process parameters.
10. Pattern classification applied to step 9, to implement quality control standard (e.g., MIL-STD-883-D, Method 2010-3).

Special aspects:
• Image scan geometry and speed.
• Processor architectures for steps 0, 1, 2, 4, 6, and 9, with shared memory between procedures 0, 1, 2, and 4, on one hand, and 6 and 9, on the other hand.

---

3. Test and qualification facilities.
4. In-process inspection.

Illustrative examples are given in Figures 38–42. The technical challenges involved are:

1. Inspection time, in view of the very large pixel sizes of the IC, ASIC, or gate array images and of alignment time.
2. Absence of absolute 3-D shape characterization rules (boundaries, slopes, edges).
3. Irregularities in the images because of emission/reflectance/field depth changes.
4. High classification error rates if the number of defect types is high.
5. Image scanning, speed, and mode.
6. Fast feedback to, or comparison with, the CAD-based IC design steps (Figure 43).

With current technology (1988), it seems difficult to achieve error rates under 10% for 60 or more defect classes in less than 10 s on

Figure 38. EPROM as visible through its transparent coating, for external visual inspection.

Figure 39. EPROM as in Figure 38 at lower resolution.

30 × 20 mm substrate areas, unless a drastic reduction in image complexity is achieved by a proper imaging or sensor fusion method (Chapter 5). The latter technique is the most promising, because it allows for specialized imaging principles and features for particular defect types and circuit activation states and provides cross-validation of detection outcomes.

Microscope imagery (Chapter 2) or other imaging methods (Section 4.6) are invaluable tools to localize and characterize defects, thus allowing for a faster feedback into the manufacturing process itself in order to weed out such faults by analyzing their causes (Figure 44). Besides, image processing of these inspection results offers the tremendous capability for generation of comparison templates to the CAD layout and for faster manufacturing turnaround time to deal with changing circuit types (Figure 43).

4.13. LINKS BETWEEN INSPECTION AND FUNCTIONAL TESTING

Inspection testers detect all detectable faults in one pass, so that their detection rate is the factor (1-yield) (see Section 1.7). Functional

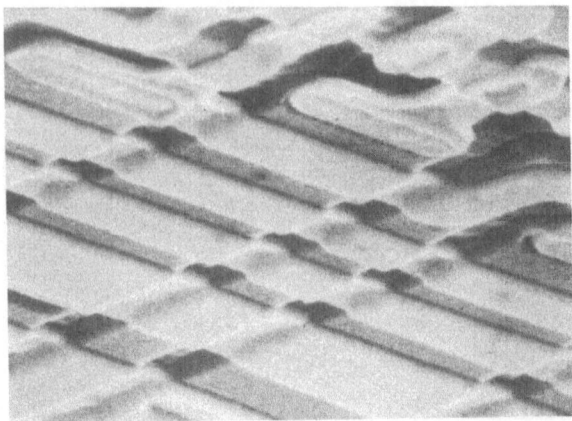

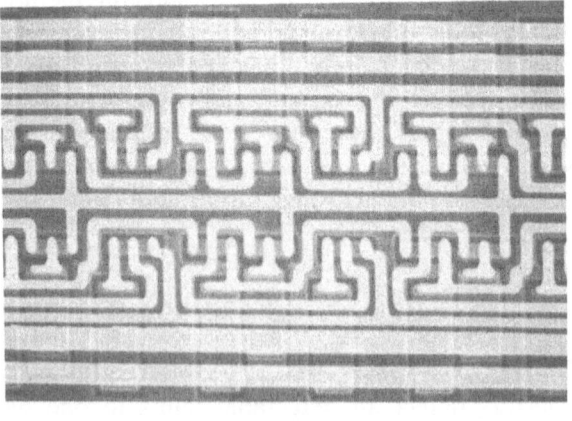

Figure 40. SEM images at various resolutions and angles.

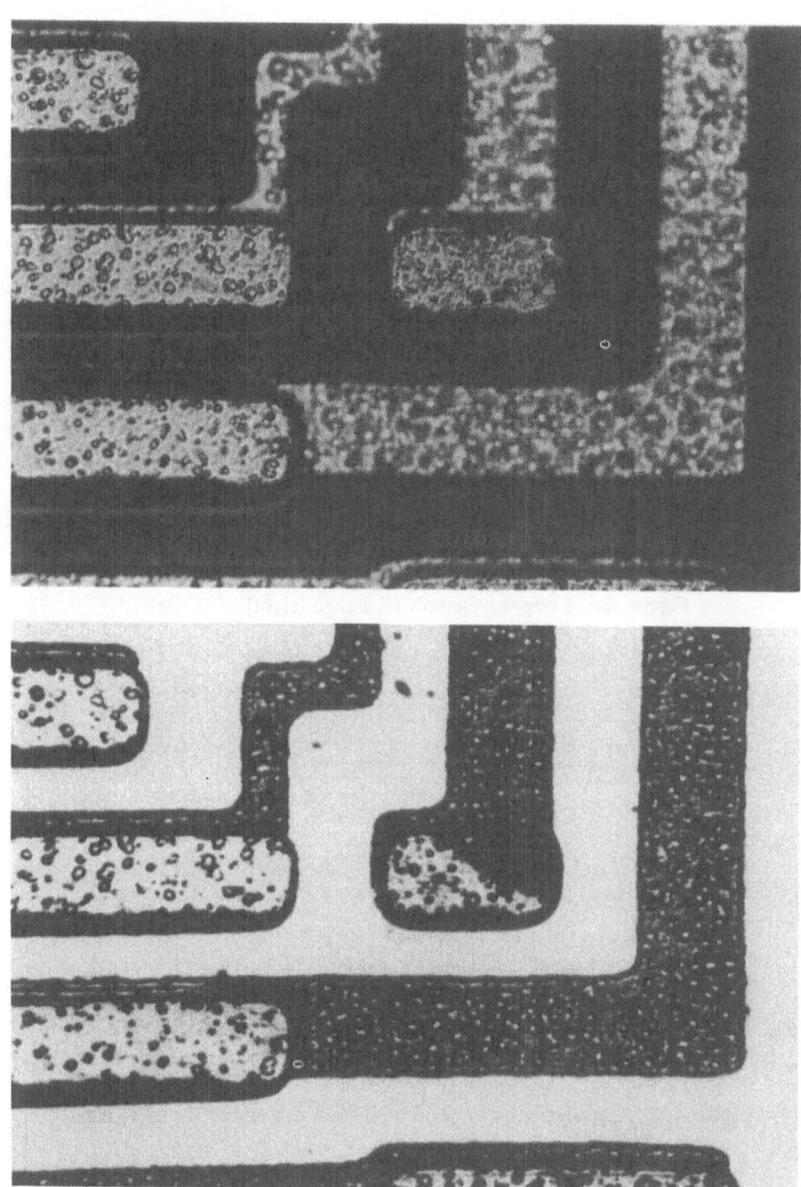

Figure 41. Laser scan microscope (top) and scanning acoustic microscope (bottom) images of the same IC at × 1160 magnification. (Courtesy Leitz.)

Figure 42. Long-wavelength IR image (right) of IC (left).

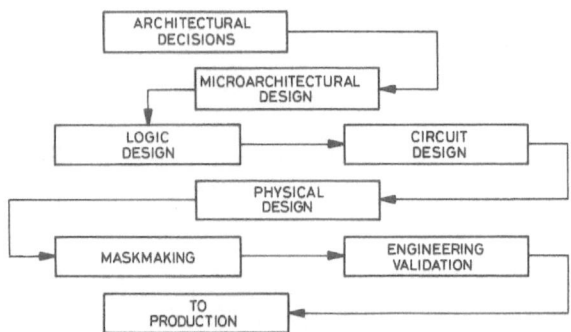

Figure 43. IC design process, where all steps are impacting inspection.

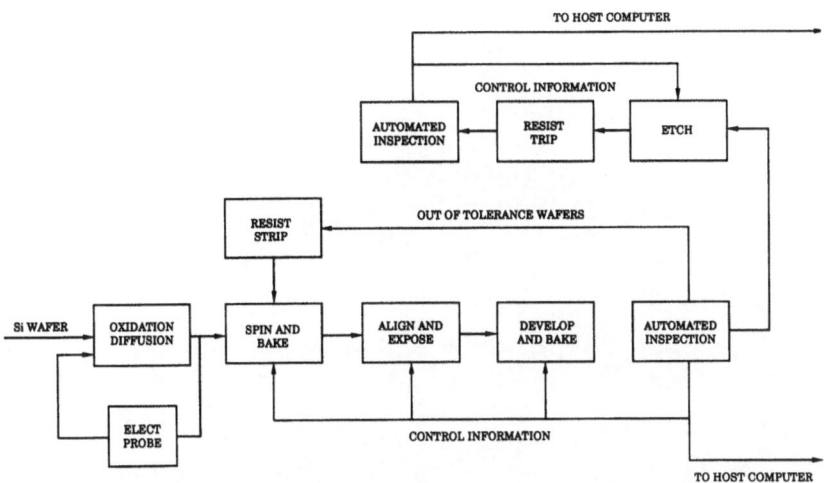

Figure 44. IC fabrication process, showing feedback from inspection to the process itself.

Table 21. Steps to Achieving the Optimum Test Strategy[a]

---

1. Estimate the technical variables (fault spectrum, process yield, production rate, product mix) based on an existing production process, or make educated assumptions about what they might be in a new process.
2. Postulate the best one or two alternative test strategies.
3. Collect data about alternative production tester characteristics. Focus particularly on intrinsic test efficiency and the cost efficiency profile for each, in the light of the variables estimated in step 1.
4. Test each strategy on paper using the technical variables of step 1 and the alternative testers in step 3. Choose the test strategy that survives this simulation best.
5. Estimate the acquistion, adaptation, and operational costs over the projected life of the product of the best alternative strategy chosen in step 4.
6. Use the cost and savings results calculated in step 5 as inputs to the particular economic model used by the company to evaluate capital equipment purchases. Compare the outputs of the model (return on investment, payback, internal rate of return, etc.) against established "hurdle rates" or other economic criteria. This comparison will give the final measure of whether or not the test strategies that look the best from the technical point of view are also acceptable by the company from the economic point of view.
7. Steps 7 and beyond depend on the outcome of step 6. If the test strategy is technically optimal and satisfies established economic criteria, the final choice is fairly clear. If the economic hurdles cannot be surmounted, then the evaluation must return to step 3 where another strategic alternative must be put through steps 4 through 6. As in any evaluation process, a "menu" like this one is only a guideline, not a prescription. But the key point remains: We probably will have to go through several iterations before settling on what strategy best satisfies both the technical realities and economic criteria.

---

[a] Ref. 41.

Figure 45.   IC test probes.

testers, on the other hand, detect one fault at a time; the factor determining the number of times devices or boards must go through the retest loop in this case is the average number of faults.

The difference between the values of the two factors will increase as the yield drops. This gap is significant in determining the optimum arrangement of inspection testers and functional testers, especially considering that the diagnosis time of functional testers tends to be much higher than test and handling time (Table 21).

Functional testers also need complex probing arrays (Figure 45), whereas inspection testers do not.

Chapter 5

Sensor Fusion for Integrated Circuit Testing

Sensor fusion[42-44] consists in combining several physical or measurement principles, in order to achieve lower false alarm and nondetection rates in the area of testing. The tremendous importance of this new approach in electronics stems from the fact that, so far, visual inspection and electrical testing have been kept separate, hence the term "integrated testing" (Figure 46). Two cases will be discussed—precap silicon IC and GaAs IC testing.

5.1. INTEGRATED TESTING OF ICs: PRINCIPLES

We will define integrated precap testing[42,43] (I) as the joint, simultaneous electrical and imaging testing on a single test station. On the other hand, sequential testing consists in performing first electrical testing (E), followed by visual inspection (V) of those ICs not yet rejected. More specifically, if we use the notation of Figure 47, the detection equations for sequential and integrated testing are defined, respectively, by:

$$H_1(S) = H_1(E), \quad \text{then } [H_0(E) \text{ and } H_1(V)] \tag{5.1}$$

$$H_1(I) = [H_1(E) \quad \text{or } H_1(V)] \tag{5.2}$$

Integrated testing will be successively evaluated in terms of IC reliability physics, test operations, and test implementation. An implementation example will be discussed in detail in Section 5.2.

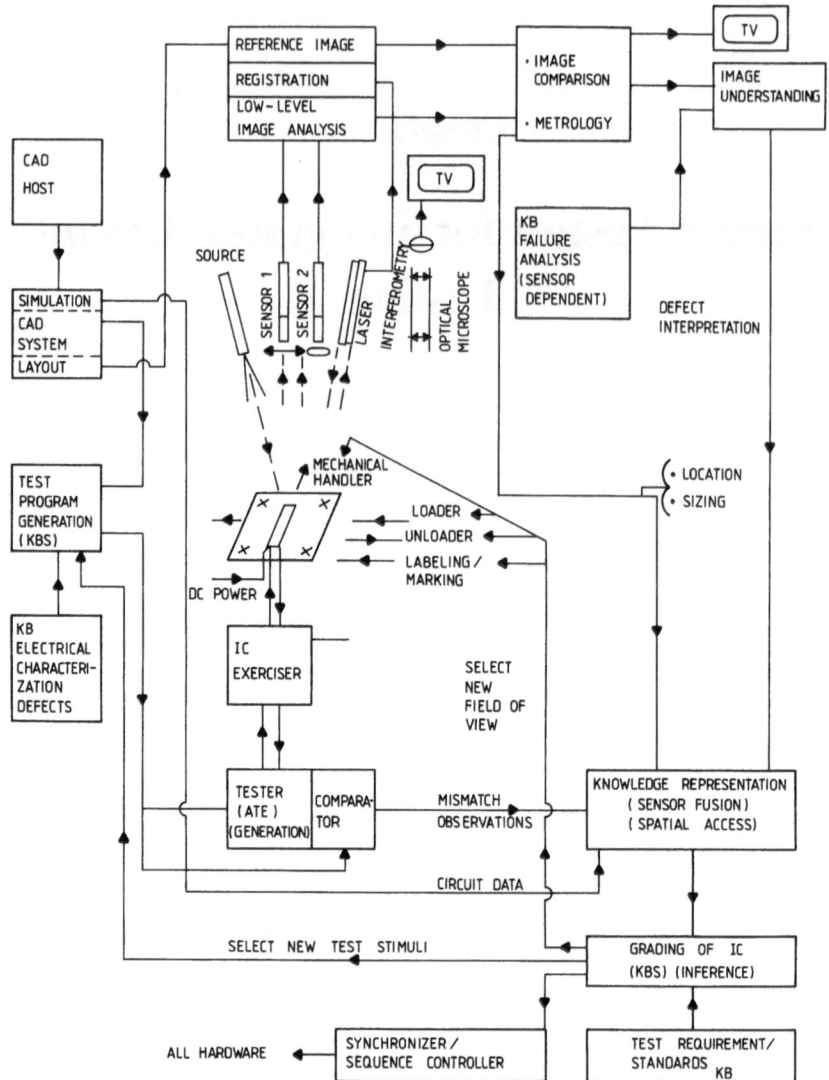

Figure 46. Sensor fusion setup for IC test and inspection. KB, knowledge base; KBS, knowledge-based system.

Among the most common defects and failure modes of ICs are the following[45–47]:

- Metallization failures caused by electromigration, corrosion, micro-cracks, scratches, or open or short circuits in general.
- Wire and die bonds (open, shorted, fatigued) and extreme leads.

Electrical testing (alone)   **Imaging inspection (alone)**

$H_0(E)$	$H_1(E)$	
$1 - \beta_E$	α_E	$\bar{H}_0(E)$
β_E	$1 - \alpha_E$	$\bar{H}_1(E)$

$H_0(V)$	$H_1(V)$	
$1 - \beta_V$	α_V	$\bar{H}_0(V)$
β_V	$1 - \alpha_V$	$\bar{H}_1(V)$

$H_0 \triangleq$ no defect
$H_1 \triangleq$ defect
$\bar{H}_0 \triangleq$ no defect detected
$\bar{H}_1 \triangleq$ defect detected

$\alpha \triangleq$ nondetection rate

$\beta \triangleq$ false alarm rate

Sequential testing (electrical followed by visual) with rejection after $\bar{H}_1(E)$

$H_0(E) \cap H_0(V)$	
$1 - \beta_S$	α_S
β_S	$1 - \alpha_S$

$1 - \alpha_S = 1 - \alpha_E - \alpha_V - \alpha_E\alpha_V + \alpha_E\beta_V + \alpha_V\beta_E$

$\beta_S = \beta_E + \beta_V - \beta_E\beta_V$

Integrated testing

$H_0(E) \cap H_0(V)$	
$1 - \beta_I$	α_I
β_I	$1 - \alpha_I$

$(1 - \alpha_I) = 1 - \alpha_E\alpha_V$

$\beta_I = \beta_E + \beta_V - \beta_E\beta_V$

Figure 47. Comparison of nondetection and false alarm rates for imaging inspection (alone) (V), electrical testing (alone) (E), sequential testing (electrical followed by visual) (S), and integrated testing with sensor fusion (I). The columns H_0 and H_1 give the true states. The rows $\bar{H}_0$ and H_1 give the results of each test procedure. E is electrical test condition; V is visual test result; α is in each case the nondetection rate, and β the false alarm rate, with subscripts giving the test procedure (E, V, S, I).

- Oxide breakdown caused by static discharge and time-dependent effects or diffusion.
- Surface defects and loose particles.
- Threshold voltage shifts caused by ionic contamination, slow trapping, contamination with mobile ions, surface charge spreading, and other dielectric failures.
- Process faults, especially oxide pinholes and diffusions, alignment errors between diffusion zones and gates, dirty photomasks, violation of design rules.
- Hot electron defects.
- Charge loss.
- Die cracks, scattering defects in wafers, flatness distortion defects.
- Packaging defects and seals.
- Thermal mismatch.

The first important point to note is that separate tests are needed for different failure modes among those listed and that no single test alone can provide acceptable false alarm and nondetection rates (see Figure 47). This remark holds also for electrical testing, even if all logical states could be tested and detected, which is usually not the case. Furthermore, even if the tests are sequenced, e.g., imaging inspection after electrical testing, the final test decisions will still reflect test inefficiency.

If we assume test outcome independence, Figure 47 shows that the false alarm rates β_S, β_I are identical, but that the detection rate $(1 - \alpha_I)$ is always greater than or equal to $(1 - \alpha_S)$. This is due to the fact that the integrated testing will detect as failed a larger share of those ICs which actually have a defect than will sequential testing, which may reject ICs after the electrical test alone on a false alarm basis without later verification by imaging inspection.

A second important point is that one may design physical integrated testing schemes which will induce interactions or dependence between the electrical and imaging test outcomes, thereby further reducing the nondetection rate α_I and the false alarm rate β_I. These schemes[48] use properties of interactions between irradiation and matter at discontinuities in the latter, when subject to guided electron conduction. On one hand, external irradiation used for imaging will generate secondary electrons also modulating the electrical test outcome. On the other hand, guided electron conduction will generate secondary emissions detectable by imaging means in a different waveband than the primary emission.

The result is to allow for another type of detection logic (see Chapter 20) supplementing Eqs. (5.1) and (5.2), and specific to integrated testing:

$$H_1(I) = R[H_0(E), H_1(E), H_0(V), H_1(V), SE, SV)] \qquad (5.3)$$

where R is a relation between all test outcomes, made context dependent through the electrical stimuli (SE) and the irradiation stimuli (SV). Within the context of Ref. 48, it has been verified experimentally that the fault coverage, i.e., the probability of detecting defects, is often increased. "Hidden faults" not revealed by electrical testing may show up in the IC emissivity images, and vice versa.

A third important point is that the integrated test station (Figure 46) eliminates any test bench transfer, at a given stage of manufacturing, thus hopefully reducing the total test time by conducting the electrical and imaging tests simultaneously. Some thermal tests may furthermore be performed by carrying out some of these electrical and imaging tests while following a specified temperature profile.

In addition, the detection logic represented by Eq. (5.3) may in some cases lead to a reduction in the length and duration of some electrical test sequences or of some imaging sequences, because of the interaction between the two.

A fourth important point concerns the image complexity, which is a limiting factor for most implementations of imaging inspection of ICs. In integrated testing, this image complexity can be reduced by selecting jointly the electrical stimuli SE and irradiation stimuli SV. By the same token, the image registration is simplified, in that the irradiation SV can be selected to drive only alignment patterns on the die, with resulting position-related electrical signals.[49,50]

5.2. IMPLEMENTATION OF INTEGRATED PRECAP TESTING OF SILICON ICs IN TWO IR BANDS

This section will describe one implementation example of the procedure discussed in Section 5.1, in order to summarize the pre-processing preceding the Algorithms Bridge-1, and Fuzzy-1.

Obviously, the pattern resolutions in the IR discussed here have little practical utility. This section serves exclusively as an illustration of a detailed step-by-step integrated testing procedure.

We will define a cell as a portion of the IC layout satisfying the following requirements:

1. Each cell is part of a standard circuit layout provided by the computer-aided design software used (e.g., standard cell libraries); the electrical probing points are generated and selected by computer-aided test design (CAT) for each such simple cell; some probing points are connected by leads to readout pins.

2. Symbolic descriptions are available for the layout of each standard cell, and any part thereof[51,52]; the layout data provide not only the interconnection topology for each cell, in the form of input–output tables of connections, but also include geometrical structures, lengths, and widths, as well as pad types and orientations.
3. Each standard cell in the library is provided with input gates; electrical stimuli can be fed in parallel into a number of similar cells through these gates.
4. The imaging frame size corresponds to adequate resolution to cover a cell.

These requirements are in line with current in-circuit testing and CAD technologies. It is not assumed that each cell can indeed be probed, as opposed to the entire circuit, but only that enough learning data are available on each cell when stimulated along with others, the number thereof being reduced as much as possible.

The steps of the integrated testing implementation are then as follows. At the design stage, two files are created for each cell and stored in a data base suitable for on-line retrieval. The first file (CAD file) gives geometrical layout parameters, in terms of features suitable for later feature matching. The second file (CAT file) gives input, probing, and output states for each electrical test stimulus pattern, SE or SV, applied to each cell. These files are retrieved by the test station once the IC type number is provided and are transferred to the test station processor in synchronization with the cell scanning sequence. The IC is, for example, placed under an infrared (IR) microscope modified in three ways. First, the detector is replaced by an array or mosaic of highest available density. Next, the optics have to be of the highest possible angular resolution. Finally, the IR emission of the IC is chopped by filters operating in two wavebands: 3–5 μm and 10–15 μm. The near-IR inspection leads to the detection of hot spots, pads, or leads under various testing stimuli patterns SE. The middle- to long-wavelength IR inspection will localize many metallization and bonding defects, again under various electrical testing stimuli patterns (Figure 42). The choice of the wavelengths, resolution, and window sizes will clearly depend on lithographic resolution, circuit density, and, not least, substrate properties. Although IR emissivity is affected by surface condition and substrate, excellent temperature resolutions can be obtained in both silicium and gallium arsenide, thus assisting failure localization when emissivity anomalies are observed while electrical testing is carried out.[53] With currently available IR optics, the resolution attainable by infrared microscopy and IR thermal wave microscopy is in the range 4–11 μm, which restricts application, at present, to medium-scale integration levels of the ICs. However, thin-film IR optics, integrated with detectors, show promise for much better geometrical resolutions, together with improved overall design of the IR imaging system.[54] Reference 54 gives in detail all distortion corrections and calibration steps to be performed.

The chip/layer/die is aligned by probe-pad test patterns containing alignment indicators. Coarse mechanical alignment prior to IC alignment may be necessary before fine alignment. To assist in the fine alignment, vertical and horizontal boundary detection takes place for the estimation of the bias and tilt angle of the IC[55] (Algorithms Reg-n and Offset-1).

Each cell in the cell scanning sequence is furthermore aligned with its long-wavelength IR image to compensate for scanning errors. This is carried out by calculating intersection points of a cross-shaped reticle and matching against the CAD file (Algorithm Offset-1). Pad positions are derived and types are assigned at each pad location from the CAD file to determine pad boundaries.

For each cell being imaged, the IC is stepped through a sequence of electrical test stimuli patterns SE retrieved from the CAT file; an IR image is acquired in each waveband for each such test pattern. The stepping of the sequence is at a speed compatible with the frame acquisition and classification rates[54] and also with substrate properties; this test stepping rate is in the range of 10 to 50 Hz. A major property of this procedure is to reduce, on the basis of localized emissivity considerations, the complexity of the IR image, by proper selection of the test stimuli from the CAT file. In particular, the multiband infrared images will not reveal too many details, as these will remain in the nonemitting background of the cell image; the complexity is thus far less than in visible domain, e-beam, or laser microscopy.

Once all test stimuli patterns for a given cell have been applied and the corresponding images have been acquired and processed, the procedure will scan the next cell, and so on, until completion.

As infrared images are blurred and have only a fairly low gray-level bandwidth,[54] the following basic preprocessing steps are applied to each image. These images are processed on a commercial real-time image processor with imbedded array and point processor.

1. Each image is quantified on four bits, with quantization profile adapted to the conductor and substrate IR emissivities (Algorithm Quant-1).
2. Thresholding and logic operations are carried out (Algorithm Thresh-n).

3. Hot spots are detected from the 3–5-μm band image through the most significant bit thresholded image.
4. The pads and center lines for the metallizations are generated in graphic overlay, using the cell alignment data and the CAD file; this creates a graph G^0 (see Figure 48; note that G^0 is not connected).
5. A thinned image of the IC cell is created, by using a specially designed window thinning algorithm (Algorithm Shrink-1); this algorithm thins the radiating metallizations and other leads (except the pads) around the nominal positions represented in the graphic overlay generated in step 4.
6. Metallization defects creating short circuits will be preserved if outside the thinning windows.
7. The thinning algorithm (Algorithm Shrink-1 or Morph-2) computes also the average metallization width at the center point of the graphic overlay, denoted $p(s)$, where s is a curvilinear abscissa in each connected part of G^0.
8. The pads are inspected by correlation with reference images for the corresponding pad types (Algorithm Reg-n).

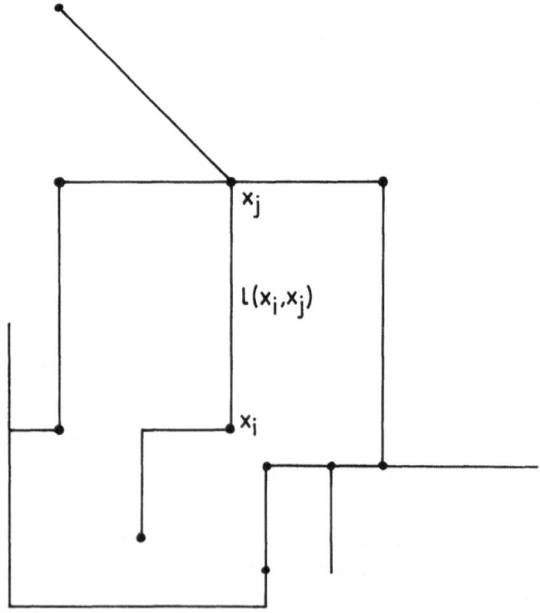

Figure 48. Labeled graph G^0 describing the IC cell, with nodes x_i, x_j and labels $l(x_i, x_j)$.

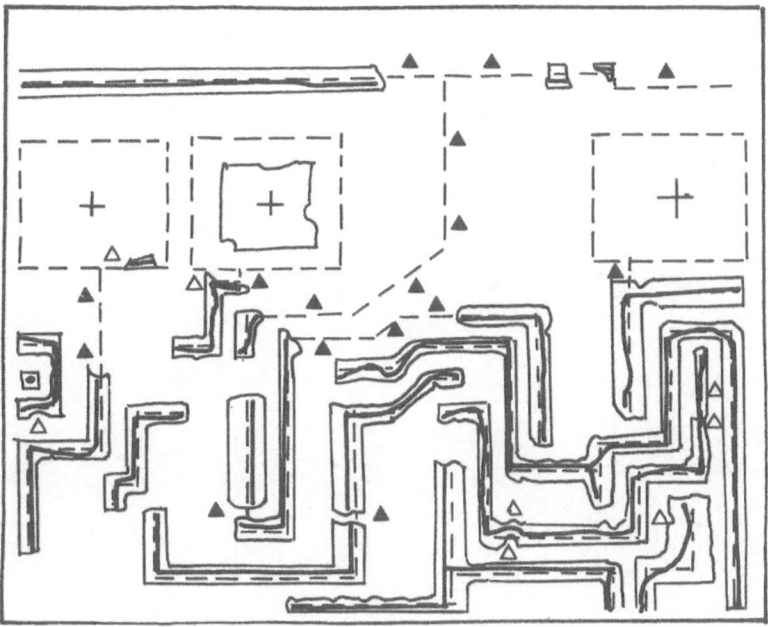

Figure 49. Roles of the different algorithms and data structures used in integrated testing of ICS: △, defects identified by Algorithm Fuzzy-1; ▲, defects identified by Algorithm Bridge-1; ———, thinned image graph G; ———, metallization boundaries (not computed in actual image); — — — CAD reference graph G^0 provided in graphic overlay.

9. The graph G^0, labeled by the attributes found in step 7, is matched to the reference cell CAD graph by Algorithm Bridge-1; this reveals all topological defects, such as opens, shorts, and too narrow patterns.

10. Finally, to further refine defect analysis when step 9 is inconclusive or to validate accepted cells versus process data, Algorithm Fuzzy-1 calculates a figure of merit for the IC cell being accepted, using a fuzzy language description of this cell with corresponding process parameters.

11. The cell is rejected if step 8 *or* 9 fails or if the figure of merit in step 10 is too small. Figures 49–52 illustrate the various cases.

The following additional remarks should be noted:

1. The above procedure may very well include the comparision of

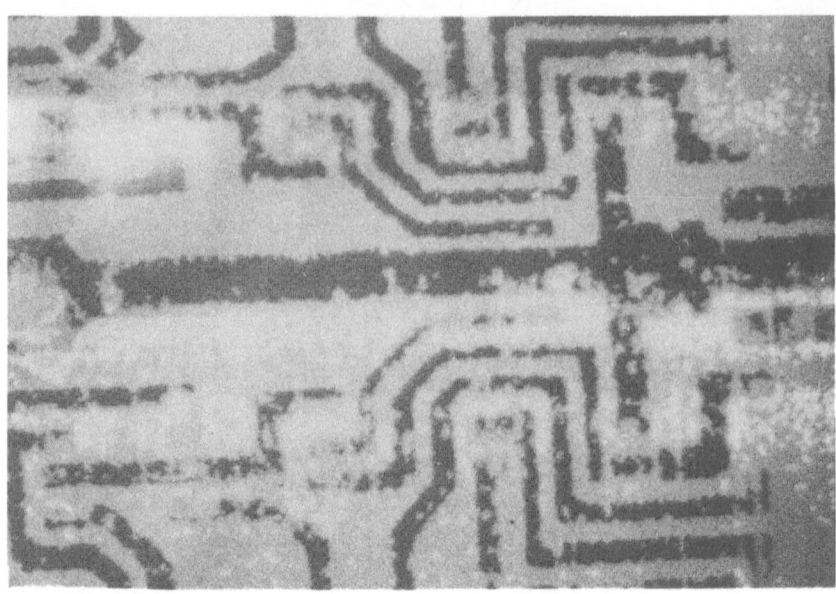

Figure 50. Digital LSM image of an IC (FET); × 200 magnification.

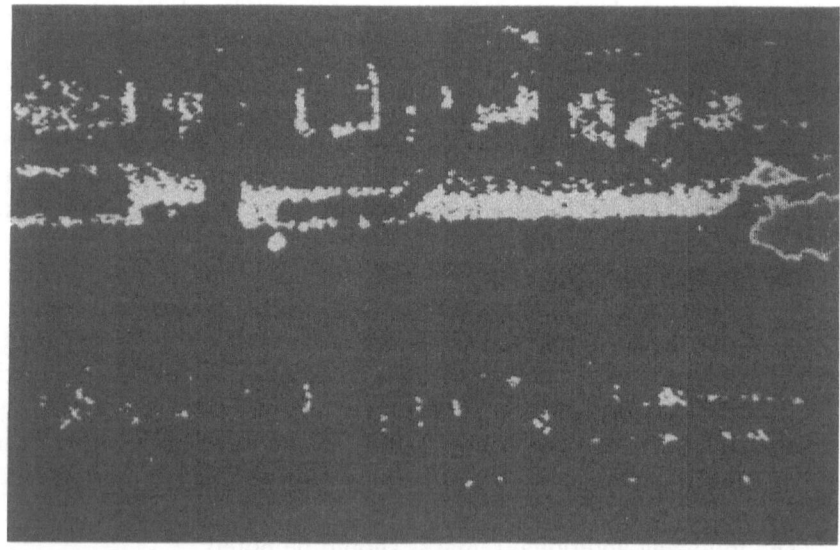

Figure 51. Defects as detected and localized after sensor fusion (LIF + OBIC), each type being pseudo-color coded; CAD layout has been removed. Correlation with etching defects, metallization defects, and electromigration for that specific stimulus SE is good.

Figure 52. Blue He–Cd laser low-resolution LSM image, with laser-induced fluorescence image in overlay highlighting the defects.

adjacent dies or chips on the same die to speed up the preprocessing (Algorithm Templ-1).

2. Such a procedure can only detect some of the defect classes listed in Section 5.1, especially metallization defects, loose particles, oxide pinholes, alignment defects, and thermal mismatch. Other defects can be found by other integrated testing procedures given in Ref. 48.

3. The irradiation stimuli effects SV include a flood UV source or, better, UV laser photoexcitation of each cell, as covered in Ref. 48.

5.3. IMAGE UNDERSTANDING OF DEFECTS IN GaAs ICs BY SENSOR FUSION

A methodology was presented in Section 5.1, called integrated testing, to combine electrical functional testing and inspection. This patented[56,57] process allows for both substantially reduced overall testing time and increased defect detection rates, as required in manufacturing (as opposed to laboratory failure analysis).

Details and experimental results will be presented in this section about a specific implementation of the above for GaAs ICs (Figure 53)

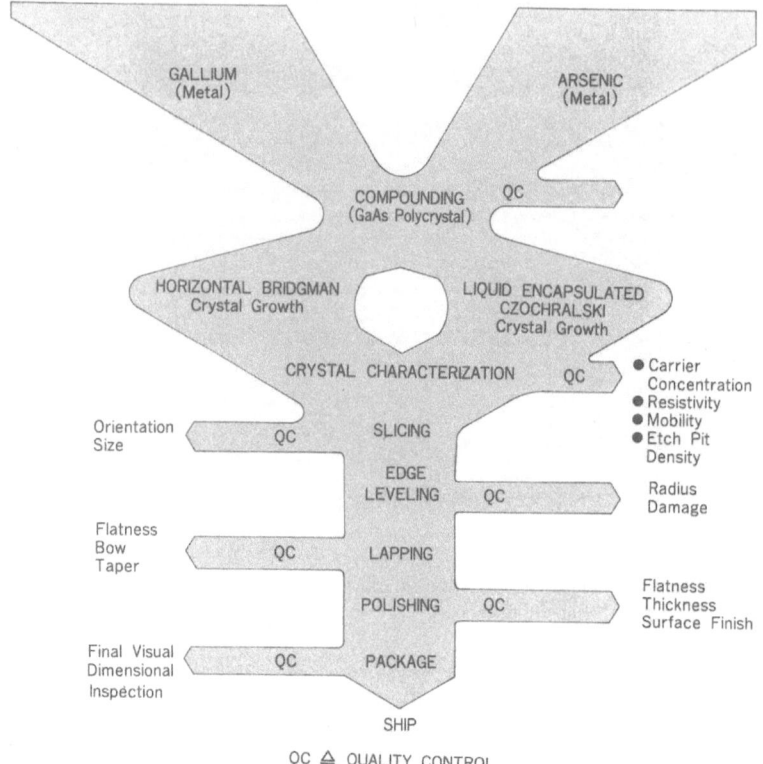

Figure 53. GaAs substrate production and inspection sequence. (Courtesy Morgan Semiconductor.)

and other III–V compounds, based on the following elements:

1. Use of laser beam-induced OBIC currents (SV), generated with a prototype of the Zeiss laser scanning microscope (LSM), with <0.6-μm resolution (see Section 2.5); substrate defects are detected by laser beam-induced fluorescence.
2. Electrical stimuli sequencer, generating controlled current excitations SE adding up to the laser beam-induced OBIC currents SV.
3. Real-time digital image processing, applied to the digital images generated by the LSM, especially for enhancement, alignment, and registration (see Section 5.3.2).
4. Artificial intelligence knowledge-based system, written in Prolog, for rule-based evaluation of the defects, comparison to the CAD

layout, and false alarm elimination, based on the results of 2 and 3 (see Chapter 8 and Algorithm Rule-1).

5.3.1. Experimental Setup

5.3.1.1. IC or Substrate Sample Preparation

In view of the demands of manufacturing quality control applications, especially fast loading/inspection + testing/unloading, only a normal clean room environment is used. There is no sample preparation, and vacuum conditions are not employed.

Once mounted under the LSM microscope, each circuit is electrically connected to four probes or a socket; one of these probes feeds electrical test stimuli (SE) generated by a functional tester, while the other probe is for readout of the signal response (S). The last two probes are for the power supply.

5.3.1.2. Sensor

A prototype (Zeiss) laser scanning microscope[57] is used with two lasers (He–Ne + He–Cd) and normal incident light microscopy. Imaging modes are listed in Figure 54, and the laser beam energy obtained as well as the beam intensity are given in Table 22. The main advantages for

Optical Laser	Incident	Transmitted
Incident He–Ne	+	+
Transmitted He–Cd	+	+

- Induced fluorescence (OIF/LIF)
- Polarized light
- Nomarski differential interference contrast
- Contrast inversion

Figure 54. LSM imaging modes. (The " + " indicates that such an imaging mode combination is possible.)

Table 22.   Laser Beam Energy and Intensity in the Object
Plane of Laser Scanning Microscope

Laser	Wavelength (nm)	Power (mW)	Efficiency	Maximum energy in object plane, E (mW)
He–Ne	633	5	0.04	0.2
He–Cd	442	5	0.16	0.8

Beam intensity for He–Cd laser

Objective	N.A.	Beam diameter, D (μm)	Numerical aperture, A (μm^2)	Beam intensity,[a] $I = E/A$ (W/mm^2)
×63	1.4	0.4	0.13	6000
×80	0.95	0.56	0.25	3200
×40	0.85	0.63	0.31	2500
×16	0.35	1.5	1.8	450
×4	0.1	5	20	40

[a] Values for no beam filter attenuation.

III–V compound and circuit inspection are:

- Operation in nonvacuum environments.
- High resolution compatible with IC failure analysis requirements.
- Beam penetration depth to detect subsurface defects or assess logic states.[58,59]

The frame scan period is 2–64 s for 512 × 512 pixel images obtained by analog–digital conversion of each laser-scanned point after photomultiplication. The digital image builds up in a 512 × 512 × 8 bits frame grabber under microprocessor control.

5.3.1.3. Optical Beam-Induced Currents (OBIC)

The laser beam SV induces currents locally; for a 2-s frame scan rate, the beam induction time is about 7.6 μs. An image is built in the frame grabber by quantizing these optically induced currents read out through the signal probe S. As the laser beam raster scans the surface, the quantized S values in each point build up a 512 × 512 × 8 bit OBIC image.

By selecting jointly the SE and S probe positions, several possible OBIC images can be generated.

In another development, transmission OBIC has been used, with a backside LSM reflection geometry, to induce and measure internal waveforms in GaAs ICs at high frequency.[60]

5.3.2. Digital Image Processing

First, the following three images are transferred from the frame grabber into digital image processor planes:

> Plane N: He–Cd laser-scanned incident image (with optionally polarized light).
>
> Plane $(N + 1)$: He–Cd laser-induced fluorescence (LIF) image.
>
> Plane $(N + 2)$: He–Cd laser beam-induced OBIC image, parameterized by the laser beam incident energy SV.

Next, the following basic processing steps are accomplished by software:

> *Step 1:* Contrast stretching and coherent addition of Plane $(N + 2)$ (Algorithm LUT-1).
>
> *Step 2:* Thresholding of Plane $(N + 1)$ for background and noise elimination (Algorithm Thresh-n).
>
> *Step 3:* Edge-enhancing filtering of image Plane N (Algorithm Edge-1 or -2).

Each of these steps is completed in less than 0.1 s.

5.3.3. Knowledge-Based Interpretation

A stored CAD layout is used for:

> *Step 4:* Logical XOR of Plane $(N + 1)$ and CAD (Algorithm Subtr-1); this detects the substrate defects revealed by LIF and located in the etched/processed areas of the layout.
>
> *Step 5:* Logical XOR of Plane $(N + 2)$ and CAD; this detects the currents due to shorts.
>
> *Step 6:* Logical XOR of the image in Plane N with the CAD layout (with transparent etched parts and black substrate areas); this detects open circuits and other defects processed in step 7.

Step 7: Detection by knowledge-based software written in Prolog of defects other than those detected in steps 4, 5, and 6 from all image planes N, $(N + 1)$, and $(N + 2)$. This patented process[57,61] will reduce false alarms and reduce the non-detection rates by detecting and validating defects by one of the procedures in substeps 1–6, yielding knowledge-based processing attributes (Algorithm Rule-1):

1. Step 4

Figure 55. GaP substrate image with defects, at ×240 magnification, scanned with a He–Cd laser with 50% beam attenuation, polarization filtering, and digital LUT (Algorithm LUT-1) for gray-level slicing.

    2. Step 5

    3. Step 6

    4. Response S false, given stimuli SE

    5. Matching bridges in the layout[56] (Algorithm Bridge-1)

    6. Figure of merit of the IC cell from a fuzzy description[56] (Algorithm Fuzzy-1)

Step 8: Repeat steps 1–7 for another test stimuli SE and for other IC circuit areas/cells after mechanical stepping.

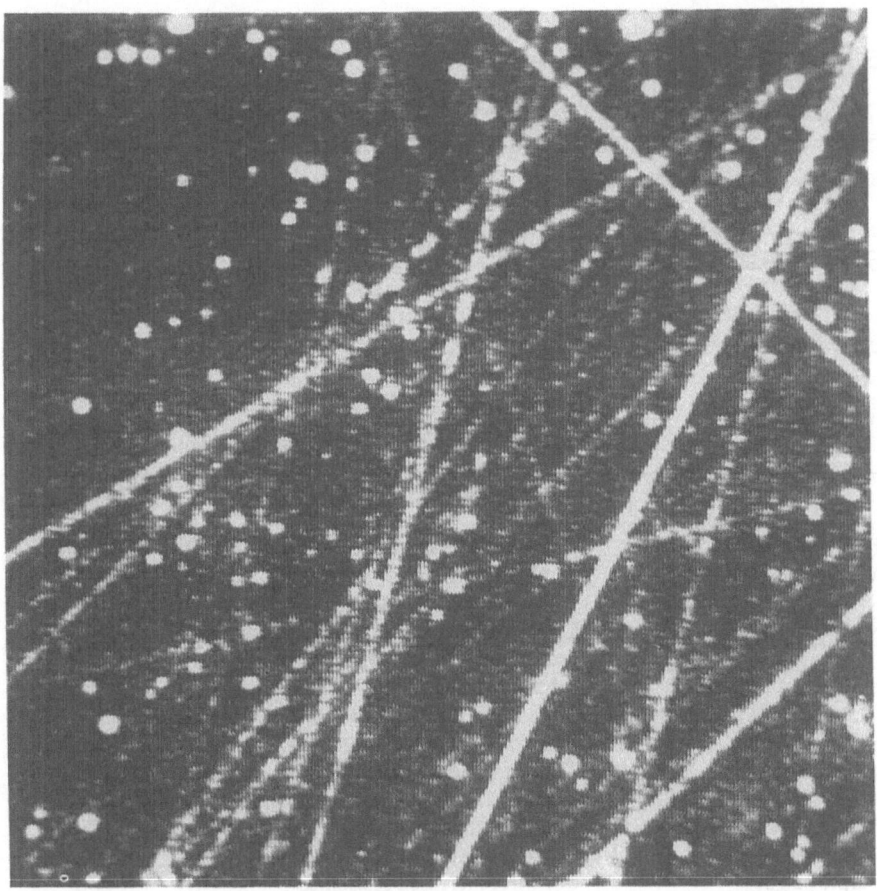

Figure 56. GaP substrate imaged by He–Cd laser-induced fluorescence (LIF), at ×180 magnification, with 9% attenuation and digital image thresholding (Algorithm Thresh-1).

5.3.4. Experimental Results for III–V Compounds and Defect Correlation

5.3.4.1. Substrate Defects

The above method (except steps 4–7) has been applied to KD substrates, KDTe substrates, and GaP substrates [dopant Zn; orientation 100; carrier concentration 2.9×10^{17}–4.4×10^{17}; mobility 73–78 cm^2/V; resistivity 0.03–0.18 Ω-cm (300 K)] (see Figures 55 and 56).

The image processing yields very fast (2 s/field of view) defect location, sizing, and counting.

5.3.4.2. GaAs IC Circuit Defects

The above method (steps 1–8) has been applied to GaAs FET circuits with 6-μm geometries. Total test and inspection time (image acquisition + image processing) for a $400 \times 400\,\mu$m window is less than 3 s.

5.3.4.3. Defect Correlation

The qualitative evaluation of the defect correlation on III–V compounds is summarized in Table 23.

Table 23. III–V Compound Defect Correlation by Sensor Fusion

Defect	Measurements	Correlation
Fixed particles	Change in V_T; short circuit of p–n junction	Medium
Moving particles	Parameter drift, leak currents; low-frequency noise	Low
Surface particles	Leak currents; capacity-induced breakdown; changes in the frequency bandwidth	Good
Surface impurities	Breakdowns because of pinholes, over-diffusion; leak currents; bad connections	Good
Substrate defects	Function loss and leak currents	Good
Metallization	Cut connection; corona breakdown; lead currents; shorts	Good
Masking and bad connections	Breakdown; function loss; short circuit	Medium
Corrosion	Cut wire, whisker leak	Low
Purple plague	Cut wire, increased resistance	Low
Electron migration	Cut wire	Medium
Parasitic fracture	Cut wire, reduced bandwidths, leak currents	Medium
Drift of parameters	Leak currents	Low
Logic states	OBIC image	Good

Chapter 6

Wafer Inspection

Wafer inspection involves wafer flatness measurements, wafer surface and probe mark inspection, and critical dimension measurement. Wafer inspection also collects data on variables essential to microlithographic process control and relates to CAD (Figure 57). The alternative imaging methods used are:

- Optical microscopy (darkfields, Nomarski/interference) (see Chapter 2).
- SEM (see Chapter 2).
- Holography[62] with on-line hologram generation of either the wafer itself or the CAD-based layout geometry, for off-line comparison.
- Confocal laser scanning (see Section 2.1.2).
- Laser scanning tomography (LST) for bulk sample imaging (see Section 2.3.3).
- On-line small-spot X-ray fluorescence (XRF).

All the above are supplemented by wafer handling and by optomechanical alignment with reference wafers. Wafer inspection is mandatory for silicon, GaAs, InP, and GaAlAs in-process wafers but extends also to resist image metrology (Figure 58). However, among wafer inspection systems, only about 40% are used for wafer-based IC fabrication, whereas 60% serve for mask inspection and repair (Chapter 7).

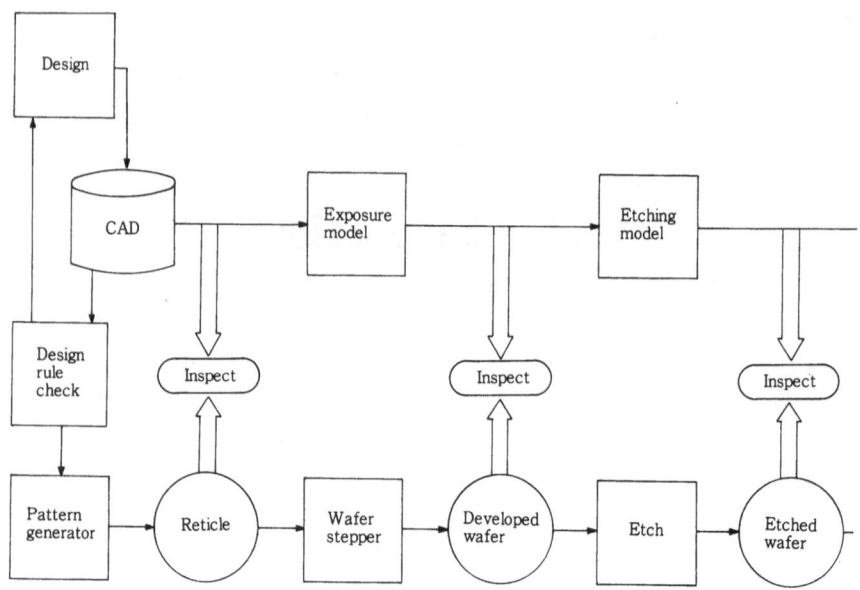

Figure 57. Inspection steps in CAD-based reticle and wafer inspection.

6.1. *X–Y–Z* STAGE OR STEP-AND-REPEAT ACCURACY

Wafer inspection, as well as other large-field-of-view inspection applications (see Chapter 10), involve mechanical scanning through $X–Y–Z$ stages or step-and-repeat systems (see Figure 59). Such systems must have the ability to represent accurately and repetitively a Cartesian coordinate system.[7] This imposes straight travel of each axis with minimum pitch, yaw, and roll, orthogonal axes, high measurement resolution and positional repeatability after software error compensation, and tight control on the operational environment and room environment. Users should always first consider X-, Y-, and Z-stage resolution performances. The Z-axis accuracy depends on automatic focusing repeatibility, which in turn depends on:

- Z-axis stage resolution.
- Depth of focus of optics.
- Sensor resolution.
- Edge detection and localization accuracy.
- Focusing software and autofocus speed.
- Illumination.

Besides wafer handling and stage movement control, the wafer

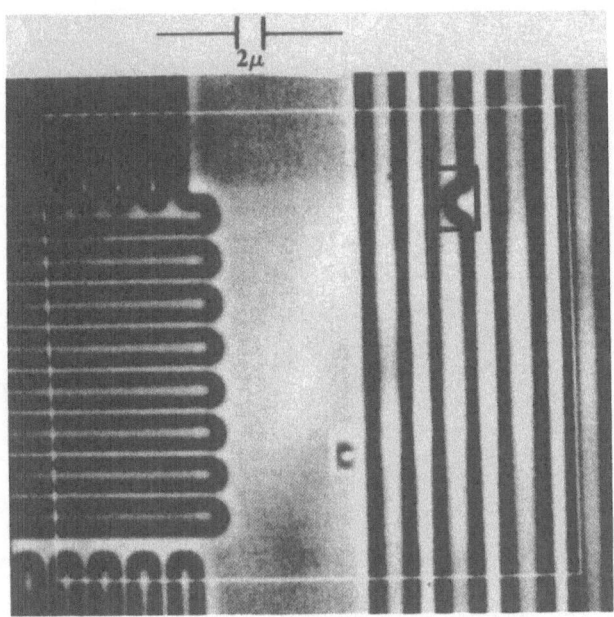

Figure 58. Photoresist protrusion and signature; one defect (top) is a curved boundary segment not allowed for by the design rule verification, and the other is a linear boundary segment, predicted but not observed (bottom). (Courtesy Cambridge Instruments.)

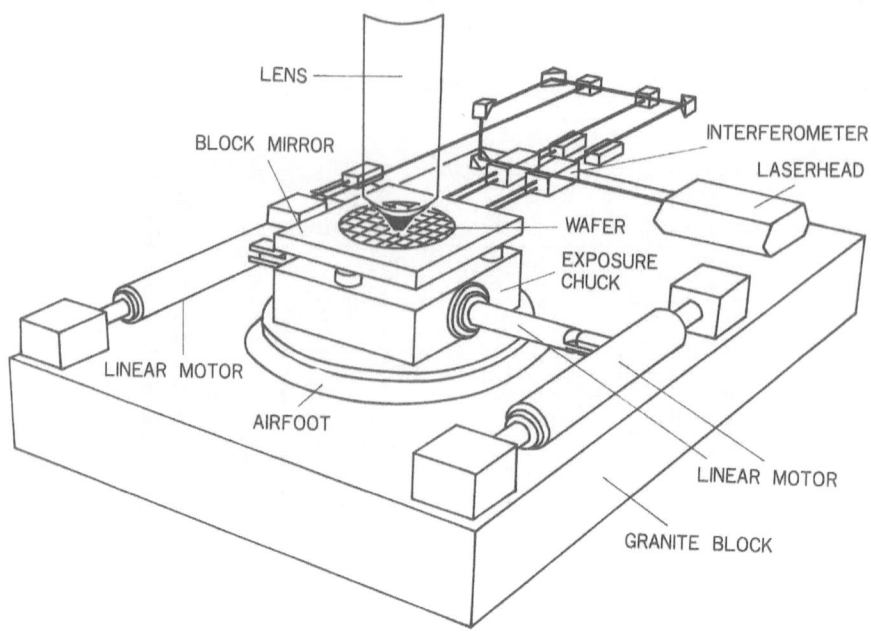

Figure 59. Step-and-repeat laser-controlled $X-Y-\theta$ stage.

inspection systems have, typically:

- Five axis eccentric stages, so that any point can be inspected without having to rotate the specimen.
- Selectable autofocus or motorized turret.
- Forward-tilting binocular eyepieces.
- Bright/darkfield.
- DIC.
- Fluorescent lighting.
- Optional: holographic gratings.

6.2. WAFER FLATNESS

The wafer flatness, F, can be calculated[63] from measurements related to the projective photolithographic tool, to the mask, and to the photoresist. Other flatness measures are given in Section 3.3.1, and an example is shown in Figure 60.

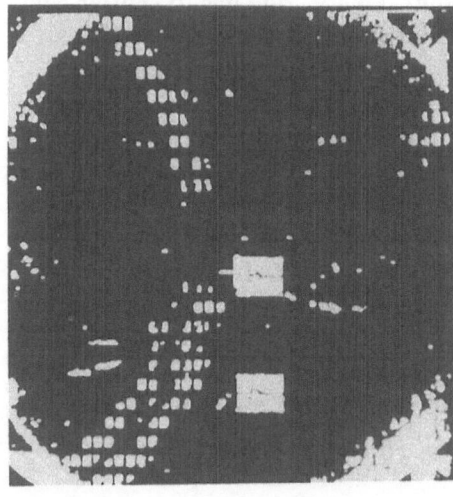

Figure 60. Wafer inspection image after processing. (Courtesy KLA Instruments Corporation.)

For full-field projection systems:

$$F = [A^2 - (B^2 + (C/n)^2 + D^2 + E^2)]^{1/2}$$

where A is the photolithographic tool depth of focus (established by the tool manufacturer from the system optics, refractive index of the medium, and wavelength of the exposure light), which is the allowable variation in distance ($+/-$) from the theoretical focal plane within which an image is satisfactorily resolved for dimensional control and image fidelity; B is the wafer-to-wafer positioning variability in the projection aligner; C is the photoresist thickness at the process level being exposed; n is the refractive index of the photoresist; D is the front surface flatness of the exposure mask; E is the product topography (maximum film thickness step at the process level being exposed); and F is the calculated wafer flatness, where flatness is defined as the maximum deviation of the elevation between any two points of the front surface while the wafer is held clamped to a flat vacuum chuck.

For step-and-repeat systems, the above equation with $D = 0$ holds.

6.3. STEP HEIGHT PROFILE MAPPING ON WAFERS

Heterogeneous structures, such as resist profiles on a wafer, are the subject of step height mapping. However, there is a phase change on reflection that differs from point to point. Hence, one has to sputter-coat these surfaces to provide a homogeneous measurement area and thus to get accurate measurements of resist profiles and linewidth features.

The technique used is phase-shift interferometry. A reference mirror in one arm of an interferometer is mounted on a piezoelectric transducer (PZT), and the reflection from this reference mirror is imaged onto a detector array. The wafer surface is placed at the focal point of the interferometric optics. A voltage is applied to the PZT, moving the reference and introducing a phase shift in the wavefront being imaged onto the detector array. Since the phase is directly proportional to the optical path difference at each point, the Z-profile of the wafer can be calculated and displayed. Taking measurements using two shorter visible wavelengths to synthesize a longer IR equivalent wavelength allows the Z-profile to be measured across steep slopes with wide phase change between adjacent pixels.[64]

6.4. NONETCHED WAFER SURFACE INSPECTION

To further improve the flatness and smoothness of wafer surfaces to maximum variations of 1.5–3 μm, surface and subsurface inspection are necessary after wafer sawing and polishing. The alternative approaches are:

- Raster or radial IR laser scan at an angle from the surface.
- Laser photon backscatter, when the beam is directed at an angle from the surface.

Both approaches allow for surface penetration by up to 5000 Å in semiconductor materials. From the measured scatter at all beam points (typically 100,000–200,000 points on a 6″ wafer), a bidirectional reflectance distribution is calculated in parts per million per steradian. The pseudo-color-coded reflectance distribution map is then produced and displayed. The low-scatter areas (usually color-coded in blue) are those with least damage and are thus suitable for high-density ICs.

In the case of wafers made of III–V compounds, it is also necessary to map the epitaxial-layer thickness and carrier concentrations.

6.5. SPATIAL SAMPLING IN NONETCHED WAFER SURFACE INSPECTION SYSTEMS

The large wafer diameters (9″) and very small etching resolutions lead to very high pixel data volumes to be inspected or to long inspection times. Therefore, in typical wafer surface inspection systems, only a few selected AOIs, or one pixel every five pixels or so, are inspected, representing only up to 1% of the wafer surface. So far, only holographic testing allows for full spatial resolution, by laser imaging the wafer under test with comparison to a reference holographic filter generated from CAD.

To determine more precisely the pixel sampling ratio, statistical acceptance sampling techniques and standard plans are used on-line in the wafer inspection system. Either simple defect detection probability calculations, with defect selection screens by size, location, or defect density (see Section 1.7), or more advanced sampling plans, including sequential acceptance plans for unattended operations,[65] are employed.

6.6. WAFER PROBE MARK INSPECTION

From rough wafer probe location information provided by mechanical systems, the vision system images each probe mark in its AOI,

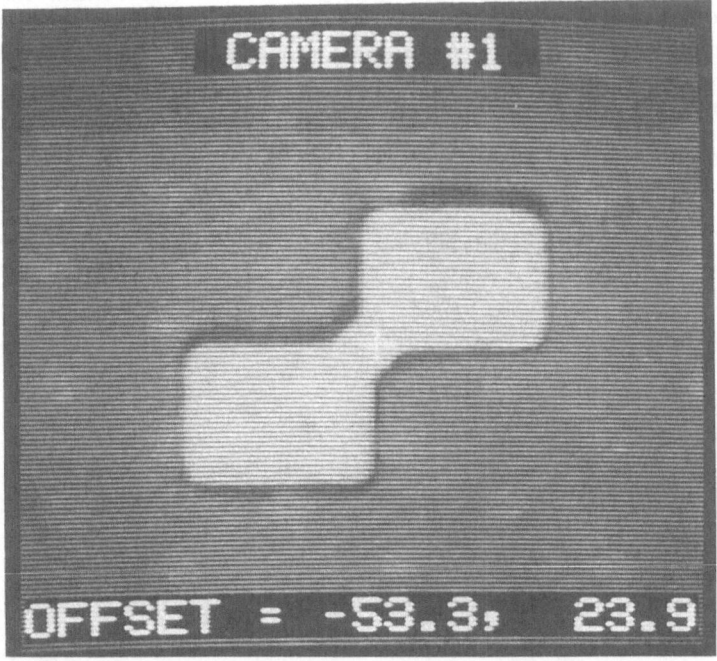

Figure 61. Alignment offset on a pattern. (Courtesy Applied Intelligent Systems.)

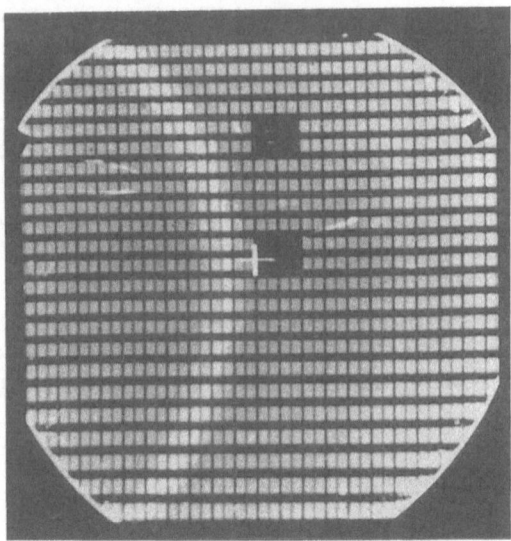

Figure 62. Etched wafer inspection. (Courtesy KLA Instruments Corporation.)

measures the distance of the encircling rectangle (Algorithm Label-3) from the edges of the bonding pad, and also compares its size with that of the nominal probe mark. An operator will be called in case of a misalignment or size deviation (Figure 61). Least-squares fit (Algorithm Geom-2 or Geom-3) can also help put each probe mark as close to the center of its bonding pad as possible.

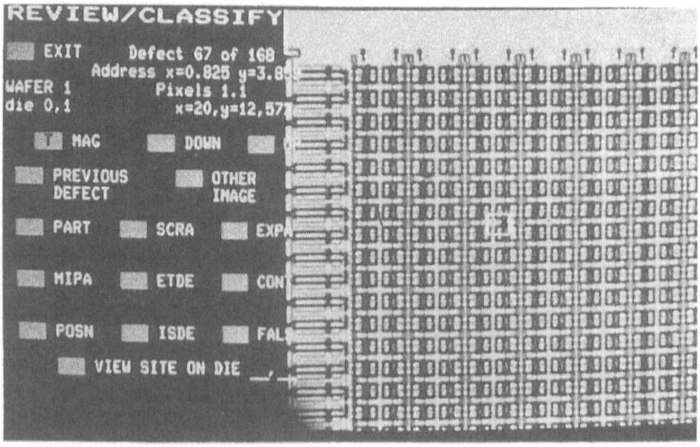

Figure 63. Wafer inspection defect review. (Courtesy KLA Instruments Corporation.)

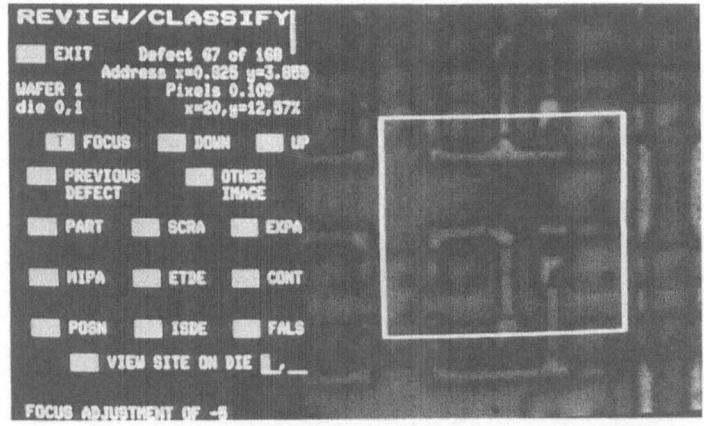

Figure 64. As Figure 63, with close-up view of the defect. (Courtesy KLA Instruments Corporation.)

Table 24. Basic Steps in Wafer Geometry Inspection from CAD

Sensor: High-resolution CCD, with scanning microscope.
Attitude: Top-down view.
Illumination: Focused or polarized beams.

0.1 CAD-1; layout generation, or generation of holographic reference.
0.2 Initial manual or mechanical alignment.
0.3 Parametric adjustments to exposure/development process.
1. Quant-1: quantization according to reflectance map.
2. LUT-1: high-pass enhancement.
3. Reg-2 or Reg-3: sequential similarity registration or twin resolution alignment to step 0.1 CAD data.
4. AOI-1: AOI processing from defect file.
5. Reg-1: correlation registration on selected features.
6. Edge-1: edge detection for linewidth detection.
7. Feature-1: attributes of connected components.
8. Morph-*n*: flaw detection with step 0.1 reference, by morphological filtering.
9. Subtr-1: subtraction for area defects in AOIs.
10. Design rule verification on step 6, using step 0.3 (see Chapter 9).
11. Label-2: position of defects in step 9.
12. Lot acceptance sampling plans for defects found.
13. Knowledge-based acceptance screens in accordance with test specifications (Algorithm Rule-1).

Special aspects:
- Wafer sorting into cassettes by step 13.
- Lithographic process control variables are data logged.
- Menu-based selection of steps 6, 7, 8, 9, and 10.
- Step-and-repeat operations from step 0.1 to step 12 with 0.1-μm laser position control, and Z-axis autofocus to 0.1 μm.
- Positional errors due to the CCD sensor velocity with respect to the wafer must be compensated for.[66]

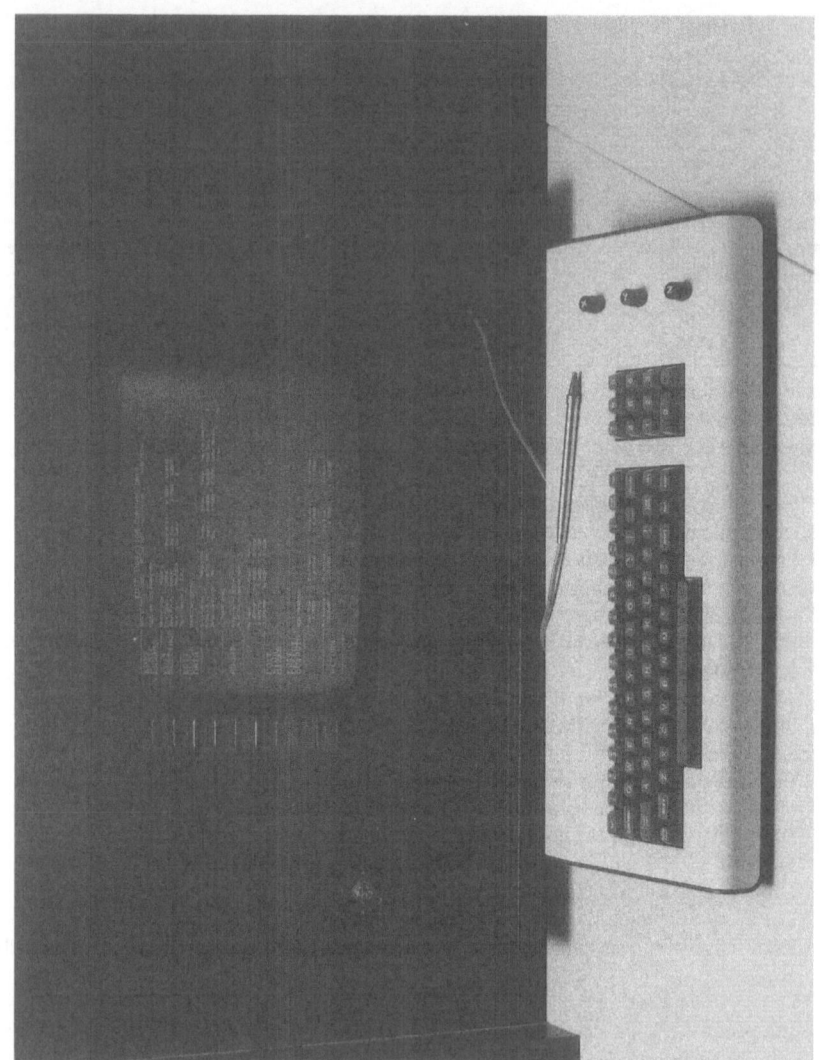

Figure 65. Wafer inspection defect review. (Courtesy Cambridge Instruments.)

6.7. CRITICAL DIMENSIONS AND MACRO DEFECT INSPECTION OF RESIST PATTERNS

The tasks considered in this section cover all variables that are critical in lithography: macro and micro defect detection, linewidth, misregistration, and area measurement (Figures 62–65) (see Table 24). They rely heavily on CAD layout data reference image generation and on registration techniques.[66] The original CAD data are often in Calma, Applicon, or other data formats (PG, *e*-beam format). Most systems are based on the principle of detecting repeating defects in corresponding locations on two dies printed from the same reticle field (Algorithm Templ-1). The primary die is inspected in its entirety and defects are catalogued in a defect file. This file is used to guide the inspection of the second die; only AOIs corresponding to entries in the defect file are inspected on the second die. Defects which correspond to defects on the primary die are flagged as repeating and are attributed to the reticle. Nonrepeating defects are considered random.

Chapter 7

Mask Repair and Inspection

7.1. MASK REPAIR

The specific pattern for each process layer of each dye is usually etched or photoplotted onto a plate known as a reticle. The reticle patterns are then reduced optically and transferred by step-and-repeat procedures to working masks or by direct step-on-wafer machines (DSW) directly onto a wafer. In any case, any flaw, defect, or error in the original reticle or mask is also transferred and will always affect final yield adversely.[67] Typical mask defects are presented in Figure 66. The types of defects are:

- Hard reticle or mask defects.
- Defects introduced during pellicle application.
- Mask contamination.
- Movement of pellicle-entrapped contamination.
- Lithography-induced defects.

In the case of ×1 reticles, an inspection and repair strategy is particularly important; because of the ×1 wafer steppers, high zero-defect specifications, and reticle-field redundancy, increased decision making is required during reticle manufacture. The maximum allowable defect sizes are 0.5 μm for 1.0-μm feature sizes, as an example.

Mask repair consists of retrieving the CAD layout, imaging the physical mask or reticle, editing its image, and activating physical

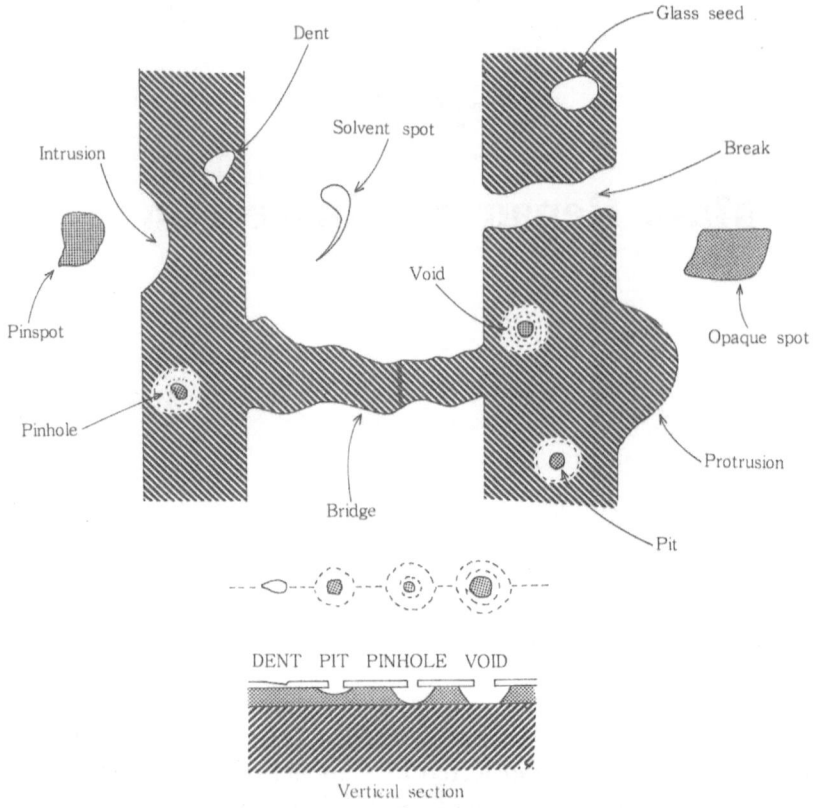

Figure 66. Typical mask defects, other than geometrical pattern errors.[68]

Table 25   Basic Steps in Mask Repair

Sensor: Video or PMT.
Attitude: Top-down view.
Illumination: See Section 1.5.2; usually with backlighting.

0.1 Retrieve and manually perform fine alignment of CAD layout.
0.2 Thresh-1: binarization of mask/reticle image.
1. Subtr-1: logical XOR with step 0.1.
2. Pseudo-1: pseudo-color coding of defects.
3. Morph-1: expansion–contraction alone, for edge cleaning and detection of missing material.
4. Morph-1: contraction–expansion alone, for detection of excess material.
5. Control repair actions based on steps 3 or 4 (see Figures 67 and 68).

Special aspects:
• Vector scan is usual for steps 3, 4, and 5.

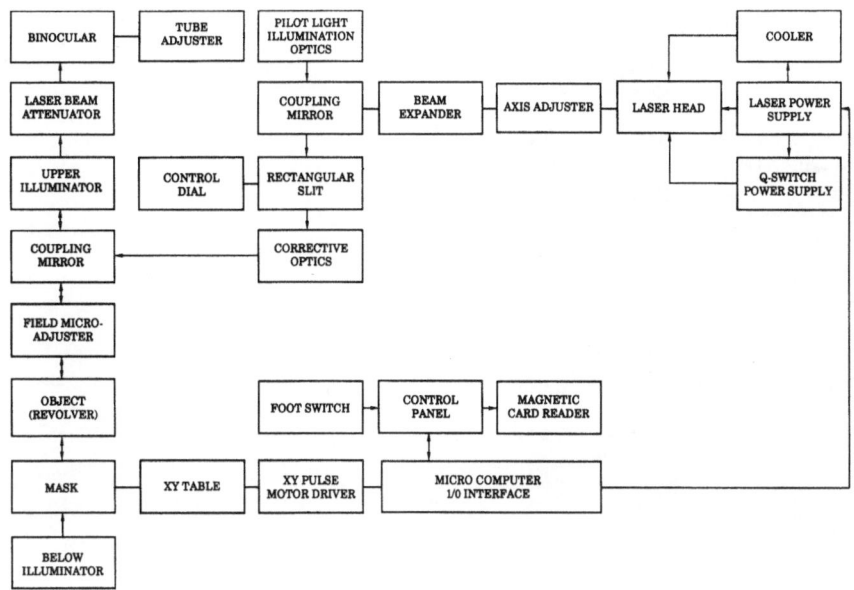

Figure 67. Block diagram of a laser mask repair system.

correction, such as removal of excess material or adding material (Table 25; Figures 67 and 68).

The imaging process uses microscopy, laser scanning, or ion scanning. The repair process involves ion-assisted deposition, ion milling, laser removal, pipettes, or dots. The defects can be as small as 0.25 μm.

An equivalent approach is used for die removal in hybrids manufacturing, where a defect file drives the repair location, to remove excess conductive epoxy.

7.2. MASK PARTICLE DETECTION OR RETICLE ERROR CHECK

Particles as small as 1 μm on both sides of masks and reticles can be detected, even after pellicles have been attached[69] (Figure 69). Such systems take advantage of the polarizing property of chrome to discriminate between the scattered light from particles and that from the mask/reticle pattern. An inspection map is produced after design rule verification image thresholding, with pseudo-color encoding of the relative particle sizes via simple pixel counting in the scanning lines (Figures 70–72). Log reports are produced (Figure 73).

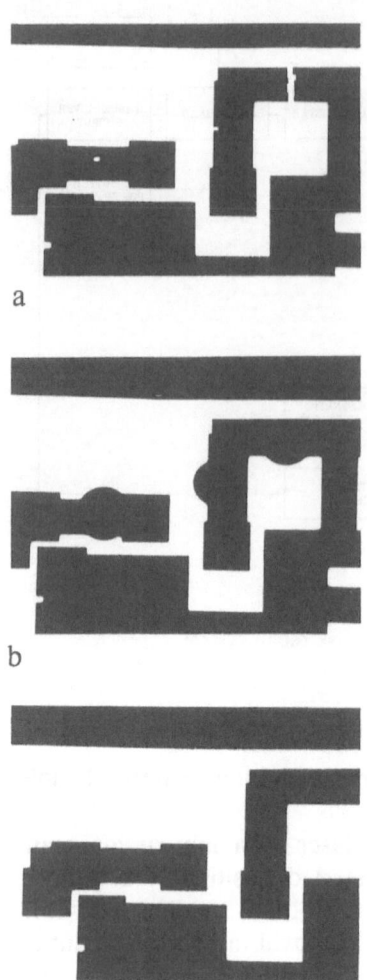

a

b

c

Figure 68. Reticle/mask repair: (a) reticle with pinhole, mouse bite, and break-in defects; (b) defect covering; (c) removal of excess material and final repair.

7.3. MASK METROLOGY

Mask metrology has many similarities to wafer metrology (see Sections 3.3 and 6.2), using almost identical vision algorithms. It deals specifically with metrology of etched antistatic ×1 masks and reticle plates (Figures 74 and 75) and metrology of X-ray mask plate structures for stability and overlay matching.

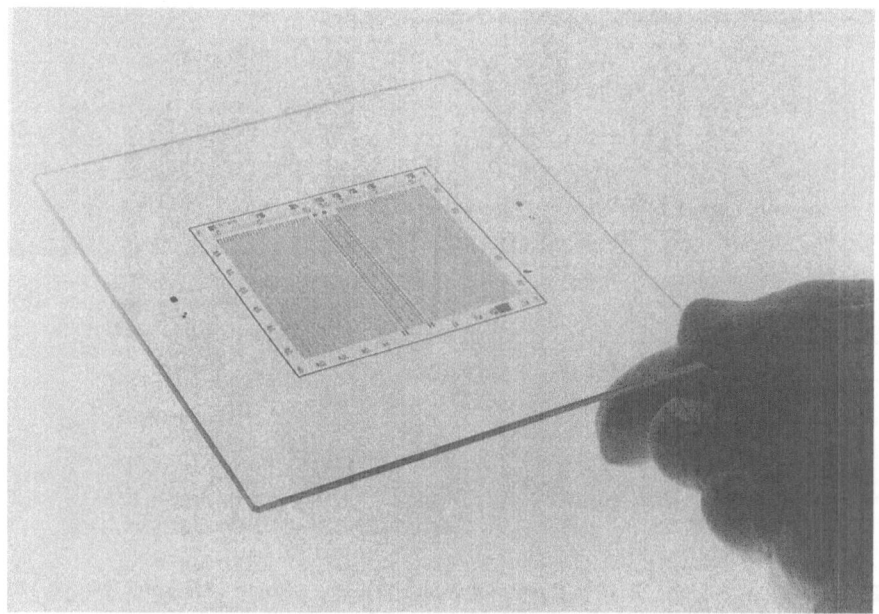

Figure 69.   Mask (mounted). (Courtesy Cambridge Instruments.)

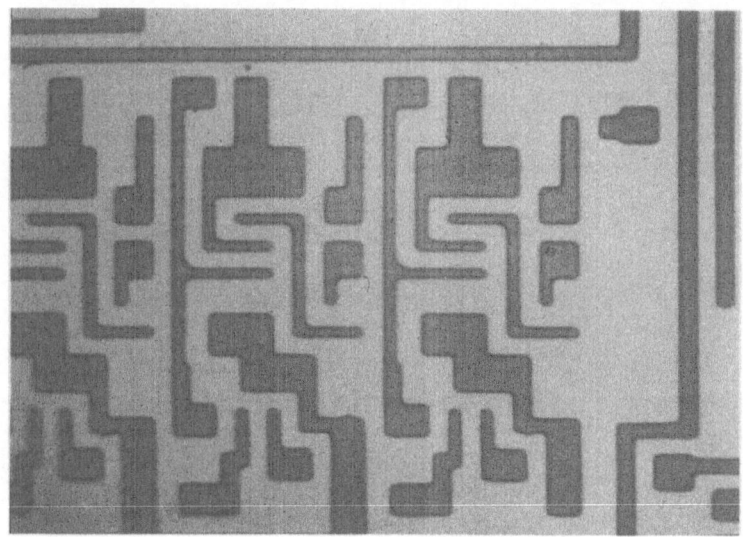

Figure 70.   Mask image. (Courtesy KLA Instruments Corporation.)

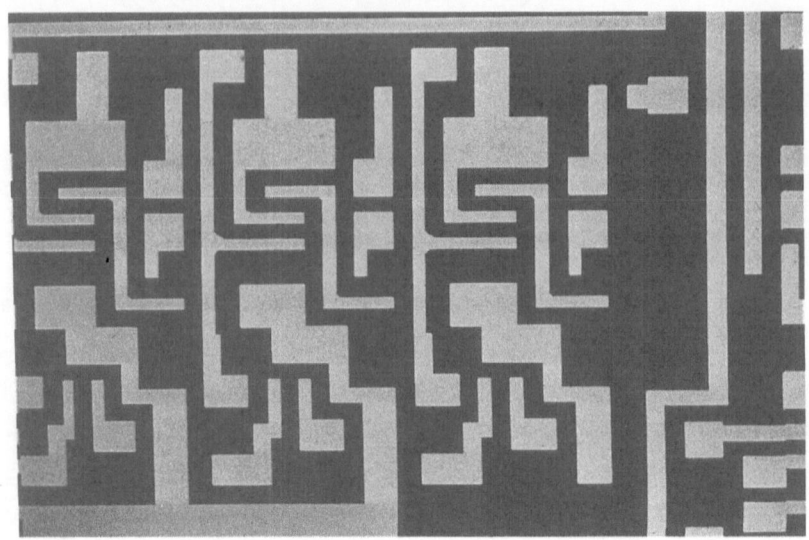

Figure 71. As Figure 70, after thresholding and two-step erosion (Algorithm Morph-1). (Courtesy KLA Instruments Corporation.)

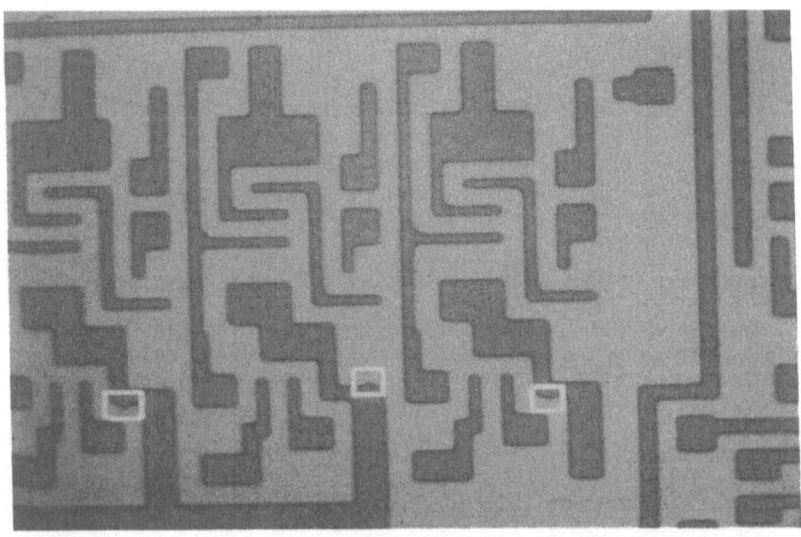

Figure 72. As Figure 70, with defects highlighted. (Courtesy KLA Instruments Corporation.)

```
****** CAMBRIDGE INSTRUMENTS LTD ------ CHIPCHECK PATTERN INSPECTION SYSTEM ******

DATE: 09-FEB-84 TIME: 22:59:28              OPERATOR : PETE FISHWICK

DEFECT LOG:    [7,30]VERP4.LOG

PATTERN IDENT: VERP4
              filter off

SYSTEM SETTINGS:
              PATTERN FILE IDENT       : VERP4.RPD
              ALIGNMENT MARKS FILE     : USERDEFINED
              OUTPUT FILE              : VERP4.LOG
              OUTPUT FORMATS           : PRINTER


DEFECT SETTINGS - USERDEFINED
                          MINIMUM PINSPOT          0.40 microns
                          MINIMUM EXTN             0.40 microns
                          MINIMUM BRIDGE           0.40 microns
                          MINIMUM PINHOLE          0.40 microns
                          MINIMUM RATBITE          0.40 microns
                          MINIMUM BREAK            0.40 microns
                          MINIMUM DIRT             0.40 microns

RUN STATUS :
              SYSTEM INITIALISED
              INSPECTION


DEFECT LOG:
  DEFECT  X FIELD  Y FIELD  X COORD  Y COORD  AIDA        LENGTH     BREADTH      AREA
    NO      NO       NO      (mm)     (mm)     CATEGORY   (microns)  (microns)  (sq.microns)
  ------  -------  -------  -------  -------   --------   ---------  ---------  -----------

     1      0.       0.      -0.188   0.185    PINHOLE      1.60       0.90        1.44
     2      1.       0.      -0.568   0.184    PINHOLE      1.60       0.90        1.44
     3      2.       0.      -0.946   0.189    PINHOLE      1.60       0.90        1.44
     4      3.       0.      -1.327   0.187    PINHOLE      1.60       0.90        1.44
     5      4.       0.      -1.707   0.185    PINHOLE      0.80       0.80        0.64
     6      6.       0.      -2.084   0.189    PINHOLE      1.20       0.80        0.96
     7      7.       0.      -2.465   0.187    PINHOLE      0.80       0.80        0.64
     8      9.       0.      -3.302   0.014    PINHOLE     10.80       1.85       20.00
     9      9.       0.      -3.294   0.021    PINHOLE      4.80       1.83        8.80
    10      0.       2.      -0.186   0.574    PINSPOT      2.00       1.12        2.24
    67      2.      13.      -0.758   3.612    PINSPOT      0.80       0.80        0.64
    68      2.      13.      -0.946   3.613    BRIDGE       5.60       1.46        8.16
    69      1.      13.      -0.566   3.610    BRIDGE       5.60       1.37        7.68
    70      0.      13.      -0.186   3.611    BRIDGE       5.60       1.60        8.96


NUMBER OF DEFECTS :    70                    DEFECTS BY TYPE:
DEFECT DENSITY    :    0.50E+03 per sq cm    MISSING           EXCESS        OTHER
INSPECTION TIME   :    2.78 mins             PINHOLE : 16  PINSPOT : 20  DIRT    : 0
AIDA TIME         :    19.85 mins            RATBITE : 2   EXTN    : 15
OVERALL TIME      :    32.12 mins            BREAK   : 7   BRIDGE  : 10
MACHINE ERRORS    :    0.                    MISSGEOM : 0  XS.GEOM : 0
```

Figure 73. Defect analysis log. (Courtesy Cambridge Instruments, Chipcheck system.)

7.4. DUAL-BEAM MASK INSPECTION

As mentioned above, a simple mask inspection procedure is to compare directly one die pattern with another on the same mask or to compare a die with a defect-free pattern generated from CAD. A similar approach is to scan two dies by a dual-beam laser scanner and compare the signals after registration error compensation (Fig. 76). Spatial image filtering (Algorithm Edge-1) then smoothes out insignificant differences, and the speed can be enhanced.

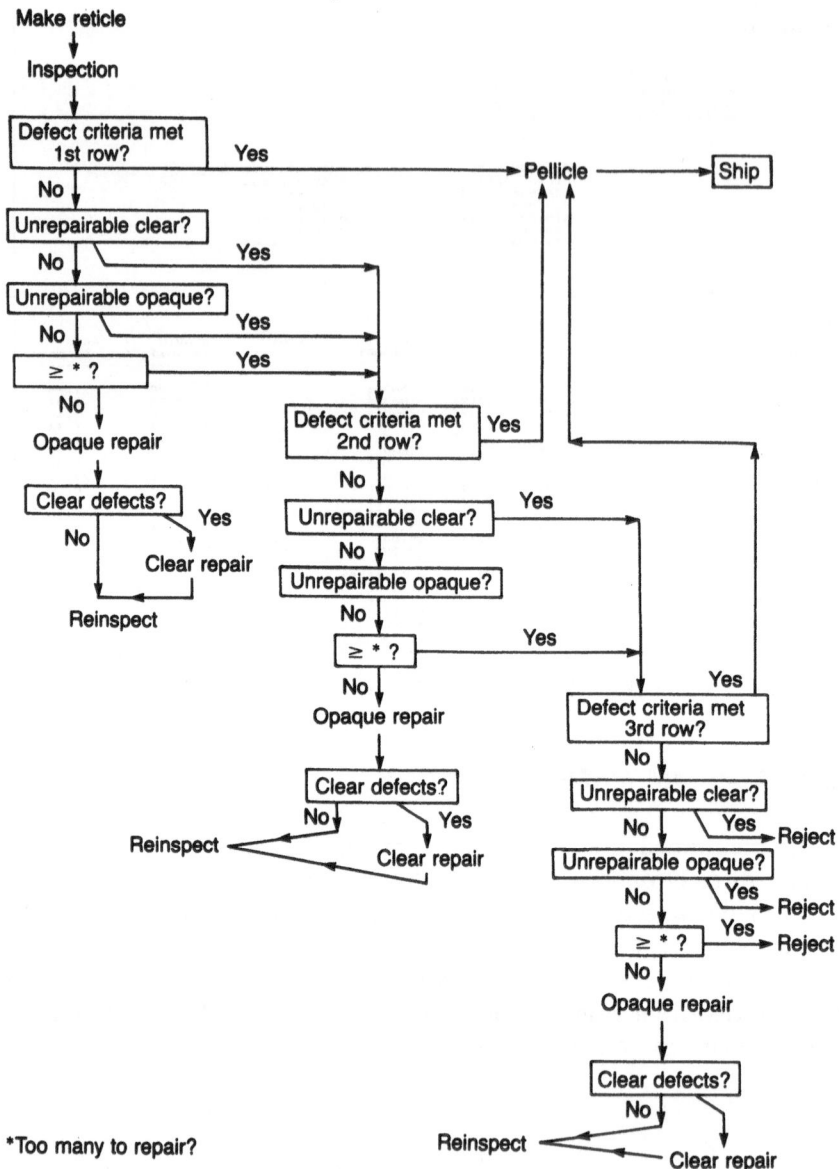

Figure 74.   Decision tree for ×1 recticle inspection (Algorithm Class-1).[70]

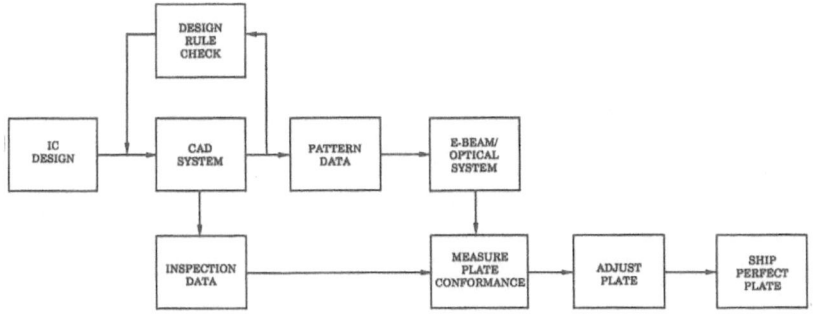

Figure 75. Photomasking with measurement and vision control of photomask plates.

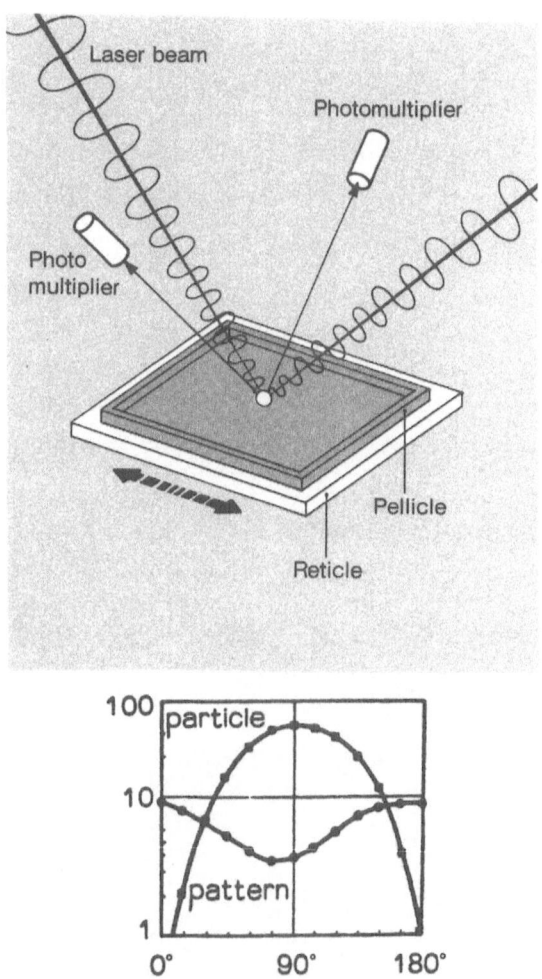

Figure 76. Mask/reticle particle detection with linearly S-polarized laser beam scans at inclination β; β is given on the horizontal axis, and the detector output (in %) on the vertical axis.

Chapter 8

Knowledge-Based Processing

8.1. PRINCIPLE OF KNOWLEDGE-BASED PROCESSING

Whereas numerical or image processing relies essentially on the manipulation and calculation of numbers, knowledge-based processing uses principles from artificial intelligence to manipulate and evaluate symbolic information, which is typically of a more qualitative nature (Figure 77). It is not the purpose of this book to present knowledge-based processing in detail; to this end, see Refs. 71–75.

However, many inspection applications, as well as process control systems, do require knowledge-based processing in order to interpret defects found in images or processes, by carrying out qualitative or symbolic evaluations of the features found through numerical or image processing. This will be exemplified in the treatment in Chapter 9 of design rule verification, which will be presented within a knowledge-based processing framework.

8.2. ARCHITECTURE OF A KNOWLEDGE-BASED SYSTEM

A typical knowledge-based system is shown schematically in Figure 78. In the following sections, the parts of such a system are described.

8.2.1. Knowledge Base (KB)

The knowledge base, stored in structured files, is the set of qualitative-symbolic procedural elements used in the interpretation of

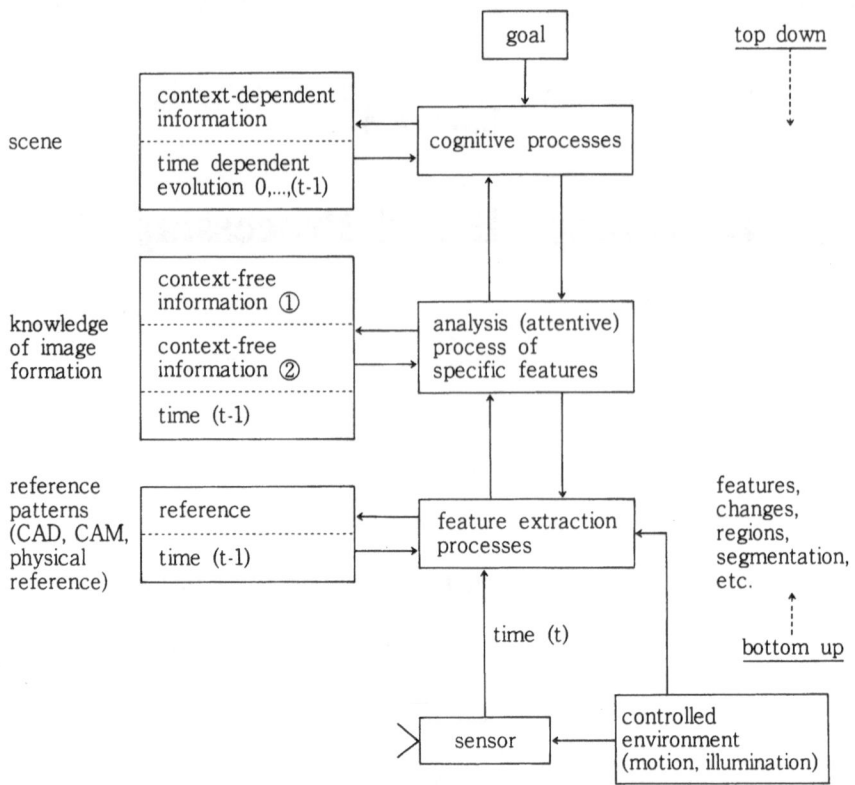

Figure 77. Scene understanding process decomposition into image processing levels, with the required knowledge base types at each level.

image or process defects (see Figure 77); those elements are structured according to some knowledge representation scheme.

8.2.2. Knowledge Representation (KR)

Examples of knowledge representation schemes, by which are encoded the contents of the knowledge base, are:

- *Production rules:*

    IF [(Condition-A) AND (Condition-B) OR (Condition-C)]

    THEN [(Conclusion-D) AND (Conclusion-E) OR (Conclusion-F)]

    A likelihood or validation probability may be attached to each rule.

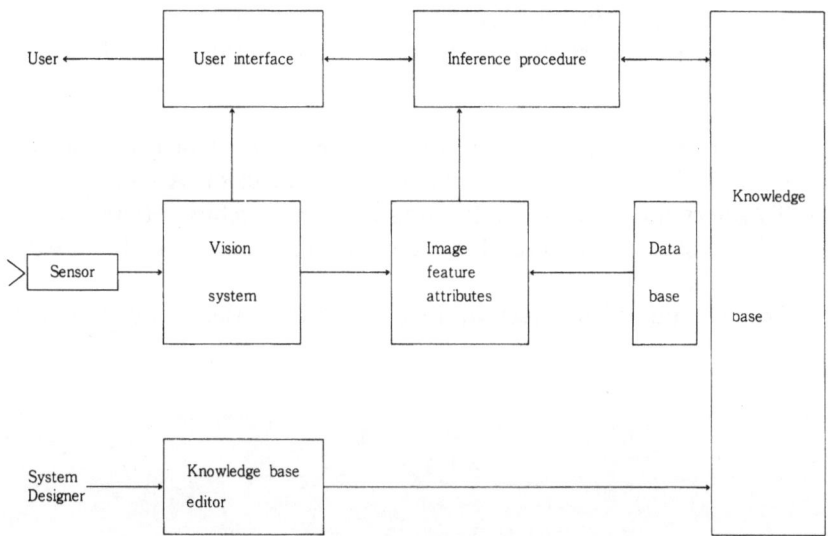

Figure 78. Knowledge-based image understanding system, showing the architecture at each level in Figure 77 and the dialogue with the user or the designer.

- *Logic predicates:* each element can only be true or false:

Conclusion-A → Condition-D Condition-B;

Conclusion-A → Condition-C;

which reads as: Conclusion-A is true only if Condition-D and Condition-B are true or if Condition-C is true.
- *Semantic network:* a procedure is described as a set of nodes in a graph, where each arc in the graph can be directed and have properties attached to it.[76,77]

In the above, all conditions apply to attributes or to earlier rules; numerical attributes are given by feature extraction (Chapter 19).

8.2.3. Inference Procedure (IE)

This designation covers a set of procedures, also called the inference engine, which implements a reasoning strategy to match a query from the outside with the knowledge base. Such a query could be, for example, to check if, or under which conditions, Conclusion-A is true. The inference procedure code is independent of the knowledge base contents.

The inference engine carries out a search and sequencing among the rules, predicates, and subgraphs in the KB to perform the following tasks:

- To detect interesting goals, that is, conditions which contain the query head, e.g., "Conclusion-A" in "Conclusion-A true?"
- To select the rules to apply using pattern matching filters, that is, select rules containing, for example, the most (or the fewest) conditions.
- To fire the rules selected, that is, assume that they hold true

Figure 79. Example of LISP processor workstation, usable for image understanding as well as knowledge-based processing. (Courtesy Texas Instruments, Explorer II.)

jointly and thus generate a new hypothetical state to be evaluated further in the search.

The inference engine can proceed in different ways depending on the query type:

- By deduction, if the query is of the type "Condition-A true?", in which case all consequences/conclusions of that condition are collected: this is called forward chaining.
- By induction, if the query is of the type "Conclusion-B true?", in which case all conditions under which that conclusion holds are collected: this is called backward chaining.

The execution of inference is fastest when implemented on special symbolic processors (Figure 79).

8.2.4. User and System Interfaces

The user interface consists of object-oriented or menu-driven software, with graphic facilities, which allows the user not only to formulate his query, but also to get explanations about how and why the results given by the inference procedure were reached. The system interface will manage all input–output with external devices, such as the image processing system, CAD system, process computer, or feature attribute data files.

8.3. ENVIRONMENTAL STRESS KNOWLEDGE BASES

A number of guidelines are available on environmental stress screening, e.g., from the Institute of Environmental Sciences (940 East Northwest Highway, Mount Prospect, Illinois 60056). These guidelines can be rewritten in production rules format [IF—THEN—(Likelihood)] (see Section 8.2.2). Rescreening, that is, subsequent screening of a device that has already successfully passed a first screening process, can be described likewise. Such knowledge bases are linked to inspection, because often screens serve to accelerate an inspected device with optical defects into early failure; as a result, the levels of rejection are interrelated.

Chapter 9

Design Rule Verification

9.1. INTRODUCTION

Design-rule-based verification looks in the image for complete paths between pads and for pads and other specified circuit nodes that should be present and connected, in order to ensure that no connections exist between nonspecified nodes (Figure 80). The corresponding electrical circuit laws are also checked. Design-rule-based verification usually relies on local image and electrical operations, process-dependent design rule knowledge bases, and dedicated pipelined parallel architecture hardware.

Most PCBs, masks, IC circuits, or thick films contain designs, such as holes, logos, serial numbers, or text, which violate design rules. Strict inspection in accordance with design rules would lead to rejection if such text areas are not masked out somehow.

Consequently, design rule checkers are actually image-based geometry and electrical functional validation procedure or spatial and electrical reasoning inference engines which specify sequences of allowable geometrical and electrical operations while excluding some zones such as text areas. The entire PCB, mask, or IC image is gauged for acceptability according to a battery of geometrical and electrical attributes, which are validated against the design rules. Some of these design rules are in turn also used in CAD design verification, which, however, manipulates data structures and not physical images or signals (Tables 26 and 27).

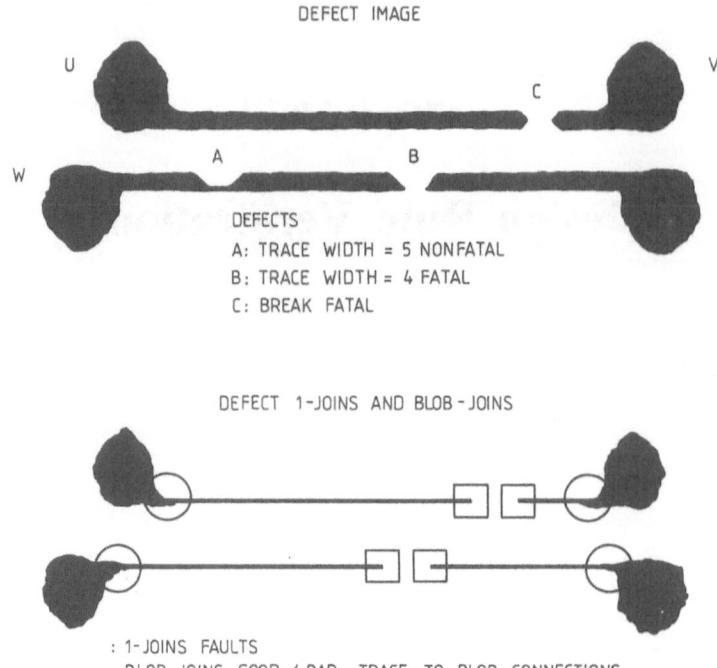

Figure 80. Design validation rules on trace width, as well as joins; the top image is the raw image, and the bottom one highlights the validated defects.

Table 26. CAD Software for IC Implementation (Verification)

Verification	Input	Output
Test program generation	Input driving signals; expected output results	Pass or fail
Graphics-driven circuit simulation	Graphic representation of IC layout; electrical parameters of material; input driving signals and output observation points	New simulation results that accurately reflect circuit behavior
Circuit continuity	Schematic or net list; graphic representation of IC layout	Exception report showing differences
Comprehensive design rule check	Design rules; graphic representation of IC layout	Graphic report of violations

Table 27. CAD Software for IC Implementation (Sizing, Mask Generation)

Mask generation	Input	Output
Logical operations on masks	Graphic representation of each mask	AND, OR, NOT, and EXOR outputs are created
Sizing	Size requirements; graphic representation of mask	Oversize, undersize, and shrink operations are performed
Pattern generation	Graphic representation of IC layout; data format required	Properly formatted tape to operate; pattern generator

9.2. LOGIC ARCHITECTURE OF DESIGN RULE VERIFICATION SYSTEMS

Inspection systems based upon design rule verification are to be considered as a special instance of knowledge-based system validation (Chapter 8), where image-based geometrical attributes and electrical properties are used to validate compliance with a process and design specification, using design rule knowledge bases (see Algorithm Rule-1).

As it turns out, these knowledge bases are best described in logic predicate form (see Section 8.2.2); the most general validation predicates would have the following structure:

$$F(V) \quad V(G) \quad V(E) \to W(G) \quad W(E) \quad W(T); \qquad (9.1)$$

$$F(I) \to F(\bigcup \{V\}); \qquad (9.2)$$

involving the basis predicates I: complete circuit, device, board, or mask; $W(X)$, $V(X)$: property X is true in window W, V; $F(V)$: window V is conditionally validated; $\bigcup \{V\}$: union of all windows V; G: geometrical properties, e.g., conditions on or value of pixel type labels and connected component shape and size; E: static electrical properties; and T: dynamic electrical properties.

The validation predicate in Eq. (9.1) says that a window and its properties are validated if some window satisfies some properties. The validation predicate in Eq. (9.2) is recursive with respect to the scanning windows and states that the circuit is validated if all windows are conditionally validated.

Specialized validation predicates can be defined by reducing the set and combination of conditions which must hold jointly.

Moreover, the validation predicates may replicate explicitly specifi-

cations from, or beyond, such standards as MIL-STD-883-C (see Sections
4.1 and 4.2).

The heart of the design validation procedure is a procedure that
examines the sequence of production rules in Eq. (9.1) as related to
windows in the image and reports any inconsistencies it discovers. This
procedure is called once for every position in the image, and the errors
are accumulated in a file showing which rules were violated, the window
containing the errors, and the coordinates of the window.

9.3. GEOMETRICAL ATTRIBUTES

The geometrical attribute extractors assume previous image preprocessing which identifies uniquely the type of each pixel (Chapters 19
and 20): conductor, substrate, resistor, dielectric, laser scribe, or text.
This defines as many image layers as there are pixel types; each pixel is
labeled by its type (see Section 18.1.3). This labeling relies essentially on
gray-level processing and simple classification rules for such gray levels in
the raw as well as filtered images. Morphomathematical procedures are
required to detect holes and specific shapes[78] such as those related to the
skeleton (Chapter 18).

Thereafter, connected components are determined in each layer,
using algorithms such as Label-n and Connect-n.

9.4. GEOMETRICAL VALIDATION PREDICATES

Geometrical validation predicates are of the form:

$$F(V) \quad V(G) \rightarrow W(G);$$

The relations between windows V and W are (see Algorithm Pat-1):

- Merging of windows and labels.
- Intersection of windows and labels.
- Union of windows and labels.
- Difference of layers in the same window.

The windows, in turn, can either be line segments, as related to
raster image scanning, square/rectangular windows in the image, or
polygons (each associated with a specific layer).

Typical validation predicates cover the following geometrical properties, G:

1. *Path width:* Check that all paths are wider than some minimum width.
2. *Path spacing:* Check that all paths are separated by some minimum spacing.
3. *Hole detection:* Detect all voids larger than some minimum size.
4. *Alignment with edge:* Check that the circuit is aligned with the edge of the substrate.
5. *Layer alignment:* Check that each layer of the window is aligned with the previous layers.
6. *Void areas:* Check that the total area of voids, outside text labels, does not exceed some fraction of the total area.
7. *Contacts:* Check that each contact is surrounded by at least one pixel of conductor in most directions.

The design rules are embodied in tables for predicate test outcomes, plus a single bit (valid/not valid) for $F(V)$.

To avoid false errors, connectivity analysis is required to nest properties in V and W windows. The simplest approach is to employ so-called critic predicates, which carry out connectivity analysis on a window V surrounding W, when W is reported to contain an error. This overhead is not noticeable, because the number of V windows that must be examined by the critic is quite small as compared to the number of W windows.

9.5. ELECTRICAL VALIDATION PREDICATES

Electrical validation predicates are of the form:

$$F(V) \quad V(E) \rightarrow W(G);$$

or

$$F(V) \quad V(E) \rightarrow W(E);$$

where the first type is called circuit extraction, that is, checking electrical properties from the image (see Section 2.5), whereas the second type is called functional testing (see Section 4.13).

9.5.1. Circuit Extraction

The designer should not be required to indicate where the nodes, transistors, passive components, or pads are or where the connections exist. The property E in V pertains to an unordered list of such basis elements, while accumulating in each layer the area and length/width ratio for each.

Therefore, the circuit or device is examined in, for example, raster scan order (left to right, top to bottom) looking through an L-shaped window V containing three W windows: the current cell W, the W cell to the left, and the W cell above. Algorithms Bound-2 and Label-1 are examples of such scanning. Using only this geometrical information, it is possible to follow electrical connectivity in V and to locate basis elements (see also Algorithm Bridge-1).

To specify circuit extraction rules, we now have to select a particular case as an example, e.g., transistors in NMOS technology.[79] There are four layers for which connectivity has to be followed:

M: Original metal layer.
P: Original polysilicon layer.
D: Derived layer, obtained by subtracting polysilicon from diffusion wires.
T: Derived layer, formed by intersecting diffusion layers and polysilicon.

The validation rules (Algorithm Rule-1) examine in turn the L-shaped window V for each layer and decide among the alternatives[80] (see Table 28).

After one pass has been made through the raster image, two files have been created: a file of merged-out node numbers and a file of transistor pieces. Due to the merging process, some of the node numbers in the transistor pieces may have been subsequently merged into other node numbers. Therefore, an additional pass is made through the transistor pieces, updating any old node numbers to their final value. Next, the pieces are sorted by their T numbers, bringing all of the pieces with the same transistor gate region together. Node numbers that were never merged and that do not appear in the list of transistors are usually text nodes or layout errors.

9.5.2. Functional Testing

The basic tests ensure that each node cell W in the circuit can be potentially pulled up and down. Thus, if prior tests have performed

Table 28. Validation Rule Structure (Section 9.5.1)

IF: $V(G)$	THEN: $V(E)$
V is empty	$F(V)$ is true
The layer is present in one W, but not in the W cells to the left or above.	Upper left corner of a new electrical node for the layer has been located and is assigned a new unique node number.
The layer is present in one W and in strictly one more to the left or above.	The layer in the current cell is just an extension of the electrical node already found in the neighboring V and is therefore given the same node number.
The layer is present in all three W.	If the node number of the layer in the left W is the same as the one in the W above, then that number is also the number of the current V. If the two first node numbers are different, then two nodes that appeared distinct previously are really part of the same node; the two node numbers are merged, and the layer in the current node is assigned the merged value.
M and either P or D is present.	The nodes for the M and the P or D layers are merged, to account for vias between layers.

circuit extraction, functional testing predicates apply to the resulting V windows and are employed to probe areas in a process-dependent way. For example, transistors that connect V_{dd} and ground together are located, and, using the length and width information, all pull-up/pull-down ratios are checked against legal values.

Chapter 10

Printed Circuit Board (PCB) Inspection

10.1. TYPICAL PCB DEFECTS AND INSPECTION REQUIREMENTS

Inspection requirements for PCBs can briefly be described as follows, with further details given in Table 29:

1. Double-sided boards
   - Board size ranges: 500 × 500 mm, with circuit sizes 80 × 80 to 120 × 120 mm.
   - Inspection speed: 0.5 m²/min, including handling.
   - Resolution: 10 μm.
   - Defects: cracks, broken wires, shorts, dislocated holes, dislocated solder resist, blurred solder resist (see Figures 81–85).
2. Multilayer boards (before, and hopefully after, lamination)
   - Board size ranges: 340 × 355 to 410 × 525 mm.
   - Inspection speed: 0.5 m²/min, including handling.
   - Resolution: 10 μm, and less for cracks and whiskers.
   - Defects: layer alignment, cracks, broken wires, whisker touch, shorts, pinholes, holes larger than minimum size, alignment with edge.
   - Check on path width, path spacing.
   - Eventually, board covered by copper foil layer on top.
3. Mask fiducial inspection on layers for multilayer boards
   - Board sizes: 340 × 340 to 510 × 610 mm.

Table 29. Checklist for Electronic PCB Inspection System Evaluation[a]

Analysis type
  Direct comparison
  Stored comparison (known good pattern or CAD)
  Design rule based

Inspectable surfaces and materials
  Silver halide film
  Glassmasters
  Etched inner layers
  Diazo film
  Photoresist on copper (red, blue, green)
  Black and red oxidized copper patterns
  Drilled boards
  Tinned conductor surfaces before/after reflow
  Spurious oxide ignored? (false alarms)
  Permissible board warping
  Other conductor surfaces (Sn, Ag, Pb/Ni)
  Other base materials (polyimide, PTFE, epoxy)
  Permissible flexibles translucency

Detectable defects
  Opens
  Shorts
  Open ends
  Minimum conductor widths (variable?)
  Mouse bites
  Nicks
  Pinholes
  Spurious residues
  Annular ring variations
  Missing feature
  Dimensional variation of board geometry
  Intermixed signal and ground layers
  Drilled hole diameters
  Pad errors
  Oblique conductors
  Isolated conductors
  Conformity to specified shapes

Economics
  True inspection speed (at given resolution)
  Maximum and minimum resolution
  Maximum total inspection area
  Maximum active area
  Automated loading
  Alignment constraints
  Known good pattern learning capacity
  Price

Table 29. (*Continued*)

Economics (*Continued*)
  Delivery
  Spares inventory
  Training
  Life of cameras and illumination source

Ergonomics
  Flaw report type (video, printout, markers)
  Flaw reporting accuracy
  Electrical effect of marking
  Pad/hole connectivity listing
  Datalogging and statistical analysis
  Verification and repair stations
  Fixturing for unit under test (UUT)
  Ease of alignment for UUT
  Type of illumination source
  Camera type or types
  Number of pixels (maximum, minimum, variable?)
  Depth of focus
  Shadow effect limitations
  Simplicity of focusing and thresholding
  Necessity for changing lenses, filters, adaptors (time to execute)
  Installation services required

---

[a] Ref. 81.

- Inspection time: 0.1 s/mark.
- Number of fiducials per layer: 7.
- Resolution: 20–40 μm.
- Defects: presence and shape of fiducials.

10.2. PCB INSPECTION APPROACHES

Three alternative PCB approaches exist, each with its own merits and weaknesses:

1. The comparison technique (Table 30) uses an already visually checked PCB board, the phototooling master, or a CAD-generated layout as the reference; the board under inspection is then compared with the reference on essentially a pixel-by-pixel basis or on a boundary-by-boundary basis (Algorithm Templ-1).
2. The feature detection technique uses a battery of circuits or software, each of which is programmed to detect a specific feature (pad, track, crack) without using a reference board (see Chapter 19); thus, this is called a nonreference technique.

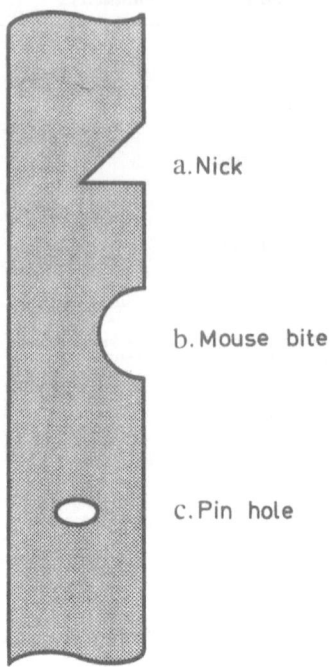

Figure 81. PCB defect geometries (I).[81]

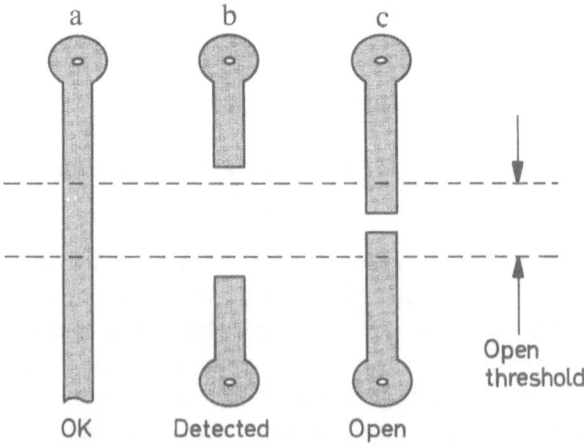

Figure 82. PCB defect geometries (II).[81]

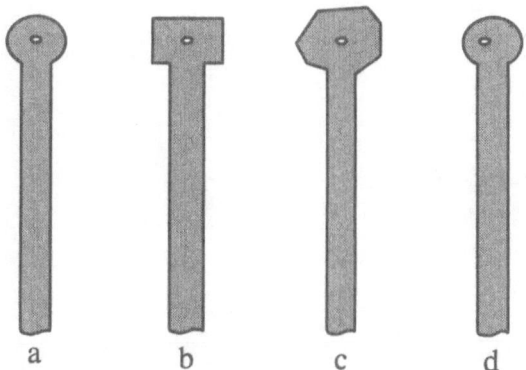

Figure 83.   PCB defect geometries (III).[81]

3. The design rule verification technique (Table 31; see Chapter 9) is concerned with determining the functionality of the board, through queries such as:

   - Is there adequate track width between each pair of pads which should be connected?
   - Are lines not ending in pads, intentionally by design, not called defects?
   - Is there adequate spacing between conductors which should be separated?
   - Are misplaced holes caused by drill walk found?
   - Does the interconnection pattern correspond with that obtained from the master board?

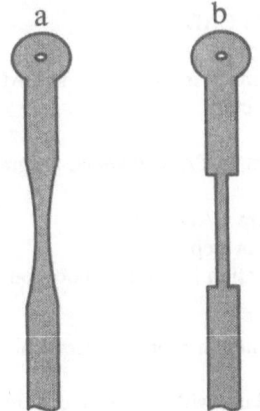

Figure 84.   PCB defect geometries (IV).[81]

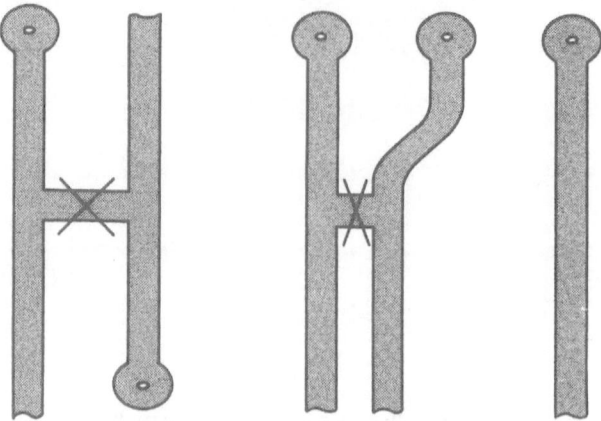

Figure 85. PCB defect geometries (V).[81]

Some systems combine the comparison technique and the design rule verification technique, double-checking each other for accuracy and thus achieving reduced false alarm rates[82] by sensor fusion as in Chapter 5.

In addition, variants may be considered which mix together these three approaches. They all share in common the need to reduce the total

Table 30. Basic Steps in PCB Inspection from CAD by the Comparison Technique

---

Sensor: Line scan video, or area CCD, or IR detector array, or PMT with fiber optic linear sensing mesh.
Attitude: Top-down view of PCB area perpendicular to its movement.
Illumination: Fixed sheet of light, or laser line scan, with diffusor behind the light sheet.

0. Generate reference layout from CAD.
1. Morph-1: expansion–contraction.
2. Subtr-1: subtraction between image from step 1 and CAD.
3. Bound-1: 3 × 3 boundary detector.
4. Bound-2: boundary tracking.
5. Region positioning with respect to CAD (manual, programmed, or Reg-1).
6. Reg-1: alignment to CAD.
7. Subtr-1: binary subtraction from CAD.
8. Feature calculation on defects in step 7.
9. Class-1: vector feature classification for defect allocation.

Special aspects:
• Steps 1, 2, 3, and 4 are performed in a single raster scan mode, with a total storage of a few scan lines.
• CAD is arranged for rapid lookup, with tolerance data for each track, pad, and hole.
• Steps 0–9 are performed in a step-and-repeat mode.

Table 31. Basic Steps in PCB Inspection from CAD by Design Rule Verification

---

Sensor: Line scan CCD.
Attitude: Top-down view of the PCB, possibly at a small angle from the vertical.
Illumination: Fixed sheet of light, LED arrays, or laser line scan.

0. CAD-1: generation of reference layout from CAD.
1. Quant-1: quantization to fix track and hole quantization levels.
2. Thresh-1: thresholding for track/nontrack discrimination.
3. Thresh-2: thresholding for hole discrimination.
4. Morph-2: erosion–dilation on step 3 to fill out pads containing holes, because the width of the annulus around each hole is less than nominal track width.
5. Label-2: positioning of the holes in step 3.
6. AOI-1: generation of AOIs for SMD pads.
7. Connect-1: generation of wiring list from step 4 (see Chapter 9).
8. Subtr-1: break or bridge detection from comparing steps 7 and 0.
9. Morph-1: on step 4 for pinholes or too narrow/wide tracks.
10. Connect-1: generation of wiring list from step 9.
11. Subtr-1: track width or crack comparison of steps 10 and 0.

Special aspects:
• Steps 4 and 9 can be replaced by specific design rule verification algorithms.
• CAD includes tolerance data for each track, pad, and hole (e.g., power supply tracks).
• Steps 0–11 are performed in a step-and-repeat mode.
• Reg-3 applies to circular holes and pads for concentricity.

---

volume of pixel data of the complete PCB down to smaller tests of features or faults. The volume of raw data can be estimated as follows. In order to image the PCB at sufficient resolution to detect significant faults, there must be at least 10 pixels across the width of a track, i.e., 0.02 mm/pixel for 0.2-mm-wide tracks. A 400 mm × 600 mm (18″ × 24″) board will generate 600 Mpixels for each side, or 150 Mbytes with four quantization levels. One such system is depicted in Figures 86 and 87.

10.3. PCB INSPECTION SYSTEM ARCHITECTURE

Visual PCB inspection systems must fulfill two opposing and equally important goals:

1. High throughput rate.
2. Versatility and intelligence.

A number of factors can limit the attainable throughput rate. Processing time comprises the major bottleneck. Although the width of vertical and horizontal traces can be measured quickly, measuring

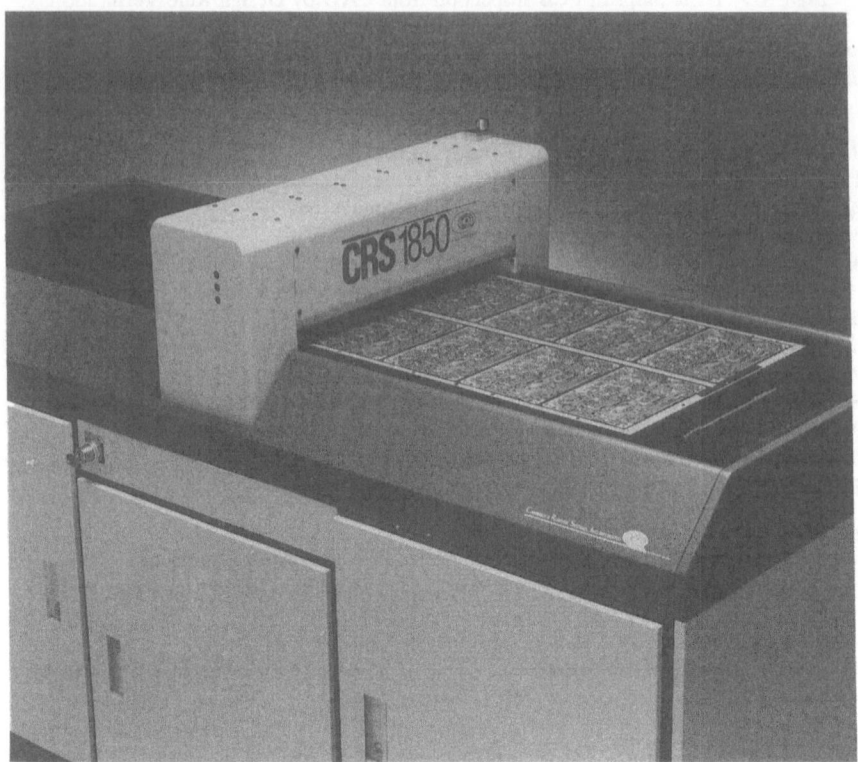

Figure 86. Example of PCB inspection system, operating on an 18″ × 24″ eight-up panel, at 0.5 mil × 0.5 mil resolution in 30 s. (Courtesy Cambridge Robotic Systems Inc.)

oblique lines regardless of orientation, recognizing pads and holes, and measuring the dimensions of the annular rings requires an automatic visual inspection system with special-purpose, dedicated hardware.

Two factors in addition to processing time impact the image acquisition process. First, the camera that images the board always operates at a certain maximal rate for frame acquisition. Second, a fundamental limit related to light intensity affects the frame acquisition rate. The faster the camera scans the board, the shorter is the exposure time (i.e., the time interval over which light is integrated). As a result, the quality of the picture declines as the speed increases. Vibrations resulting from scan rate represent still another factor limiting processing speed.

The speed of an inspection machine can be evaluated in one of two ways. In the usual method, the user measures speed in square feet of

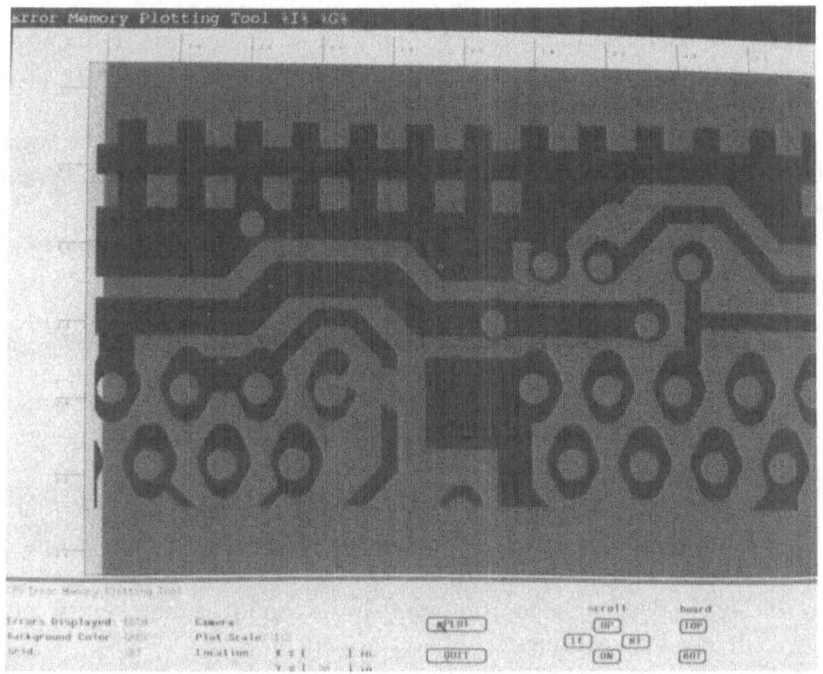

Figure 87. PCB error reporting and positioning, corresponding to Figure 86: two errors are found by camera 3 and are visible immediately to the left of the center. (Courtesy Cambridge Robotic Systems Inc.)

board area per minute. The other way, perhaps more important for evaluating the performance of an inspection system, is in pixels per second.

Each inspection system can handle pixels up a certain maximum rate. However, it may handle more square feet per minute by using larger pixels, or fewer square feet using smaller pixels and implies lower inspection rates.

One approach simply fixes the pixel size to suit the smallest features the machine will inspect. Typically, such a size falls in the range of 0.3 to 0.5 mil, resulting in a unvaryingly slow inspection speed.

The second approach allows the use of different pixel sizes for different applications of the machine. This approach offers the advantage that for applications not requiring the smallest pixel size, a larger pixel is used allowing a higher speed.

In a variable-size pixel system, two factors combine to dictate the ideal pixel size for an application: the physical size of the features to be inspected (e.g., linewidth) and the criteria and tolerances for inspection. For example, some production stages may allow a reduction in linewidth of 20%, permitting coarser resolution, while the artwork stage may allow only a 10% reduction, requiring a fine resolution.

Accordingly, a system that can maximize pixel size within the limits dictated by the combination of the application's physical feature dimensions and inspection criteria will provide the best throughput.

In assessing the speed of a PCB automatic visual inspection system, the following questions should be considered:

- Is the inspection speed constant, or does it change with the inspection requirements?
- If the speed is constant, what is it? There should be a well-defined answer in square feet per minute or in seconds per square foot.
- If the speed varies, then:
  - What inspection speeds apply for various conditions?
  - Can speed change as a function of the physical dimensions of the features on the board?
  - Does speed change continuously or among a number of pre-determined settings?
- Can the user change the inspection criteria (such as maximum allowable reduction in linewidth)? If so, does the change affect inspection time?
- Do all the system's inspection capabilities, including options, occur at the same inspection rate, or do some options require additional inspection time?

10.4. PCB ILLUMINATION

Printed circuit patterns have a vertical thickness and are not simply two-dimensional. For example, etched copper patterns comprise an unetched top surface with slope and etched side walls.

Moreover, the substrate conductor and pad areas have highly different reflectances. The specular reflectance of most substrate materials is more than a factor of 6 db below that of copper and other conductors. Variations in diffuse reflectances between substrate materials are high. The illumination must also keep fingerprints, smudges, and tarnish from causing false alarms. Examples of PCB illumination problems are given in Figures 88 and 89.

Ring-shaped optical fiber illumination sources from above give high contrast for conductor path defects. Direct illumination gives a good

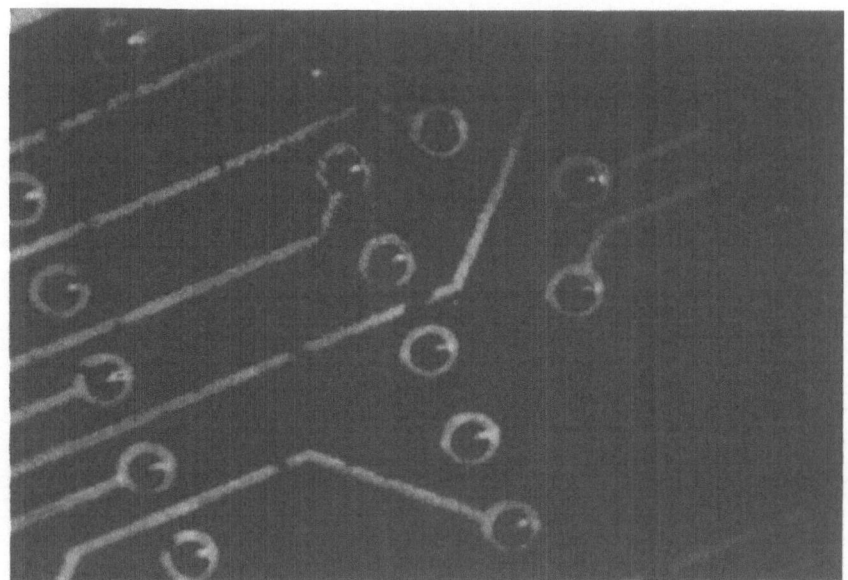

Figure 88. Example of a raw thresholded PCB digital image, with illumination problems.

contrast between most conductors and most substrates or resist materials; however, shadowing and glare are a hindrance in the case of tin–lead-plated conductors or of uneven surfaces. In all cases, uniformly diffuse illumination helps. Transillumination, or fluorescence of the laminate (when possible), with a UV or laser source gives a good contrast, for example, on epoxy–glass substrates if the sensor has a high sensitivity.[83]

Regarding optics, a long acylindrical lens looking at the laser- or otherwise illuminated sheet of light helps in getting higher spatial resolution on the PCB.

10.5. PCB ANNULAR RING INSPECTION

Measuring the annular ring on PCBs requires identifying the pad and the hole, measuring the dimensions of the annular ring surrounding the hole, and judging those dimensions in light of the inspection criteria. Criteria for the annular ring may vary from the criteria for minimum linewidth.

Two problems further complicate annular ring inspection. First, many boards contain holes of different sizes. The PCB inspection system must identify and inspect more than one hole size on a given board.

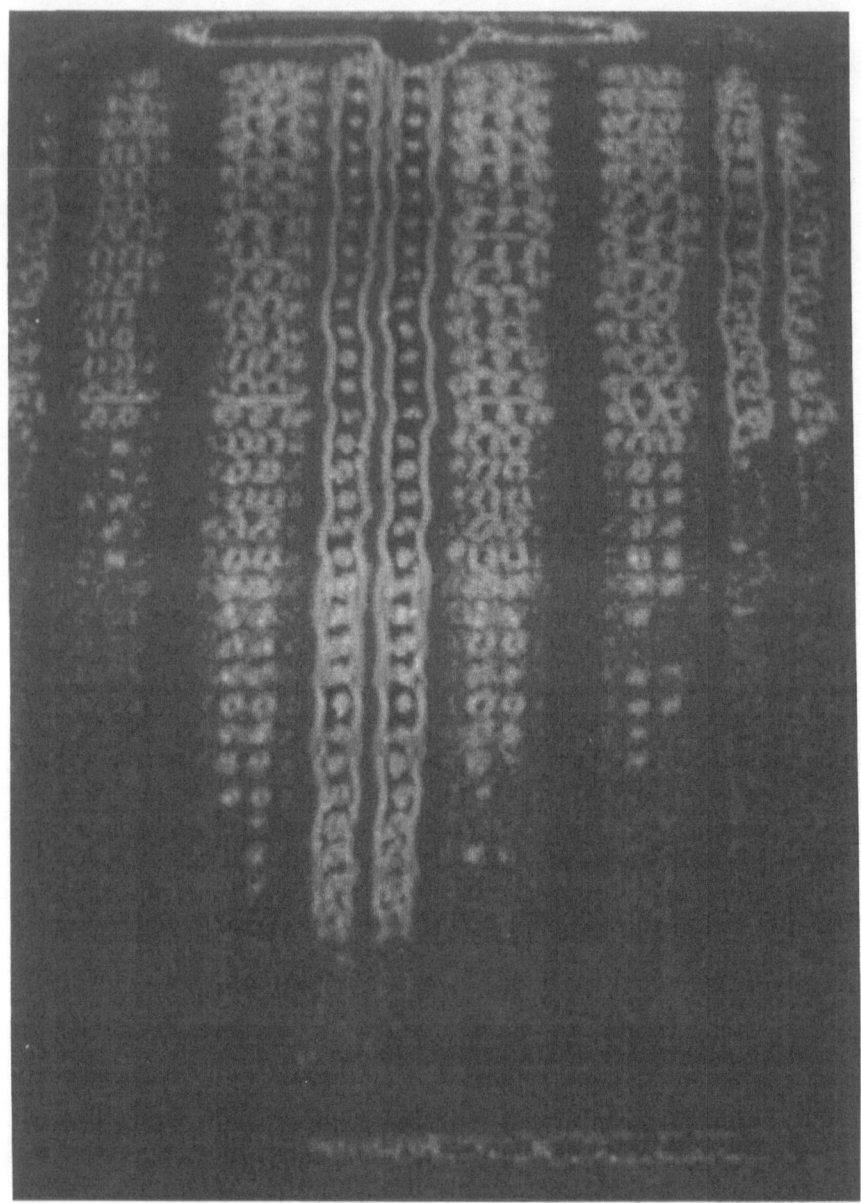

Figure 89. Example of PCB digital image after edge detection, with illumination problems.

Second, the system should recognize a case of zero annular ring and make an appropriate decision according to the inspection criteria. Some applications permit zero ring; others identify it as a defect.

Two cases illustrate the importance of this annular ring inspection capability:

- In the less usual case, minimum annular ring size exceeds the criterion for minimum linewidth. If the system cannot identify the annular ring as such, it fails to apply the correct inspection criteria and fails to identify flaws.
- In the more frequent case, linewidths exceed the minimum allowable annular ring dimensions. Sometimes, the hole can lie partially outside the pad. In this second case, if a system cannot identify pads and holes as such, then it faces a serious difficulty. Since it applies the inspection criteria indiscriminately, it will find a multitude of false alarms (perceived, but not actual, flaws) as it applies inappropriate criteria to the conductor around the hole.

As a result, systems incapable of annular ring measurement cannot inspect drilled boards, which means they cannot inspect most outer layers, double-sided boards, and inner layers with buried vias.

The following questions should be considered in undertaking annular ring inspection:

- Does the system contain a module for identifying and inspecting annular ring?
- Can the system correctly inspect annular rings when the minimum annular ring dimensions exceed the minimum conductor width dimensions?
- Can it correctly inspect annular rings when the minimum conductor width exceeds the minimum annular ring dimensions? If so, would it signal incorrect (nonexistent) flaws?
- Can it identify holes lying partially outside the pads?
- How many different holes sizes can it inspect on the same board?
- Does the inspection of annular rings reduce the overall inspection rate? If so, by what percent?

10.6. PCB CONDUCTOR WIDTH MEASUREMENT

An important part of PCB inspection is the measurement of conductor widths and of spacing widths, which are difficult measurements at high inspection rates, except under certain restrictions. For example, if all conductors must run horizontally or vertically, then widths can easily be measured perpendicularly to line direction. However, if conductor lines do not hold to particular angular orientations, then the proper measurement of widths becomes considerably more complicated.

To circumvent the difficulty of measuring nonperpendicular lines, a system can forgo true line measurements and compute the ratio of conductor width to space width instead. This technique offers simplicity and insensitivity to the orientation of the lines. On the other hand, it requires a constant conductor-to-space ratio everywhere on the board.

A better approach performs measurements of the dimensions of lines (that is, of conductors and spaces). This technique requires a measurement, at every point along a line, of the line's width as measured perpendicularly to the line's local orientation. A system that can perform such measurements on lines of every orientation can be said to perform an angle-independent true measurement of the lines.

The following questions should be considered in undertaking line-width measurement:

- Does the system measure the actual width of conductor traces and of spacings, or does the system only compute a conductor-to-space ratio?
- If it actually measures their width, does the system assume that the lines run at any particular angles, or is it angle independent?
- Is the linewidth violation always set at a constant fraction of the nominal linewidth (e.g., a 20% reduction width) or can the user specify it?
- Can the user set conductor-width violations and spacing-width violations independently of one another?
- Does linewidth measurement affect inspection speed?

10.7. VISION FEEDBACK FOR PCB DRILLING

The drilling of PCBs under visual control is a problem in which both the mechanics and the scene analysis are relatively simple. The solution to the problem could, however, have genuine practical application.

In research laboratories and other establishments in which PCBs are made in very short runs of small numbers of boards, the boards are often drilled by hand using a single-spindle drilling machine.

The operator views the board through the eyepiece, which contains a cross hair indicating the position of the drill. To drill a hole, he moves the board until the point to be drilled coincides with the cross hair, and then actuates the drill. Equipment has been built to automate this process and replace the human operator by a computer vision system. The board is viewed, via a half-silvered mirror, by a TV camera which is interfaced to a computer. The video signal is sampled with a maximum resolution of 300 × 400 pixels over the field of view of the camera, and each pixel is digitized to 5 bits.

A simple circular feature search provides the drilling position.

10.8. OTHER PCB INSPECTION SENSORS

Whereas most inspection systems today use linear or matrix CCD sensor arrays, they still suffer from significant nondetection or false detection rates and surface condition (copper, solder, etc.).

New sensors, typically line scanned, are being developed to give the full conductive/copper pattern geometry with 15-μm resolution, and possibly electrical characterization measurements. Such sensors include infrared linear arrays in the long-wavelength domain as well as sensors based on X-ray, microwave, and laser scanning.

10.9. INFRARED PCB INSPECTION

In infrared PCB inspection, an IR mosaic detector, or a scanning thermal radiometer, operating in the 8–14-μm range, is used instead of video imaging, and the comparison technique is essentially followed. If proper calibration and PCB insulation are provided, this technique produces, through lookup table setup (Algorithm LUT-1), the color-coded temperature profile map of the board. Because of noise, image averaging is necessary (Algorithm Buff-1).

The major types of defects found by IR PCB inspection are (Table 32):

- Shorts for PCBs in multilayer boards.
- Poor hybrid bonding quality, as revealed by thermal testing.
- Bad components and boards, as revealed by testing the backside of boards and hybrids.
- Opens in plated-through holes.
- Poor control of heat sinks.

Spatial emissivity can be compensated for, as explained in Section 1.5.4, through area-of-interest-based processing (see Figure 90 for an example).[84]

Table 32. PCB and Hybrid Faults Detected through Infrared Imaging

Item	Fault
Capacitor	Current leakage
	MTBF (short life)
	Reversed polarity
	Incorrect value
Chassis assembly	Hot spots
Conductor	Nicks, voids
	Cold solder joints
	Fatigued wires
	Lifted lands
Conformal coating	Pin holds, voids
	Scratches
Diode	Incorrect forward resistance
	Reversed polarity
False circuit	Electrical noise
	Faulty component
Heat sink	Poor thermal contact
	Poor design
Hybrid circuit/components	MTBF (short life)
	Inoperative TTL gates
	Hidden flaws or cracks
	Absence of power supply inputs
	Poor chip-to-substrate bond quality
Operational amplifier	Poor chip thermal bond
	Open circuit
Plated-through holes	Missing plate through (open)
	Incomplete plate through
Resistor	Incorrect value
	Poor electrical contact
	Nonhomogeneous composition
Resistor, thin film	Poor design
	Manufacturing resistance variations
Relay	Inoperable switch
	Missing drive power
Solder joint	Voids
	Cold solder
Transformer	Open winding
	Shorted windings
Transistor	Thermal runaway
	Poor chip thermal bond
	Open circuit
Wirewrap plane, multilayer PCB	Short circuits

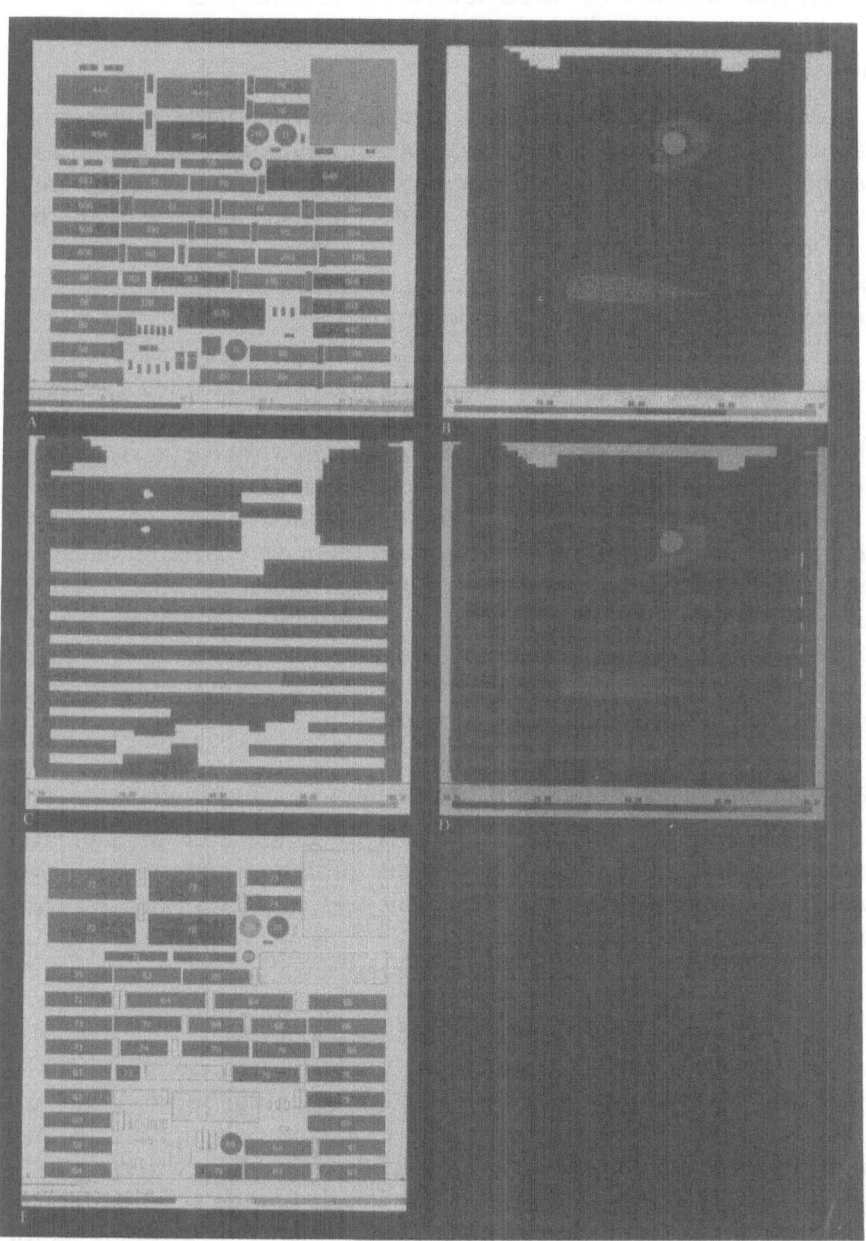

Figure 90. AOI-based processing, for thermal PCB imaging[84]: (A) gives component types, (B) the substrate layout, (E) the IR emissivity corrections, (D) the thermal image after thresholding, emissivity corrections, and pseudo-color coding, and (B) the same as (D) after subtracting the background (C). (A color version of this figure can be found at the end of Part I, after Chapter 12, facing page 204.)

10.10. INSPECTION OF MULTILAYER SUBSTRATES

10.10.1. Introduction

Multilayer, as a term, refers to methods of printed circuit board fabrication and cofired ceramic circuit manufacture, as well as processing methods for circuit fabrication based on thin- and thick-film technologies when several overlaid layers are used.[64] In each instance, very real problems are electrical shorts through insulation layers and between adjacent conductor traces, as well as open circuits in the interconnection pattern due to manufacturing (or design) errors.

Typical defects relate either to the printing process (blocked screens, smears during handling) or to other anomalies which often cannot be visually detected through several layers (short, misalignment, continuity problems). Misregistration between layers can also occur, leading to requirements for inspection prior to and after lamination.

At the same time, because of the labor-intensive efforts required in their fabrication, multilayer structures are the most expensive substrates upon which to add components. Substrates are made multilayer for circuitry densification; thus, multilayer substrates contain more components and are more valuable. It is therefore of utmost value to detect and eliminate defective substrates prior to assembly. Most multilayers are also too complicated for unaided inspection.

10.10.2. Multilayer Inspection Approaches

In view of the difficulties, vision is often not used in multilayer inspection; instead, pattern graphics are projected and superimposed onto the layers, with manual alignment and visual checking by comparison with the graphical overlay.

Typical vision-based approaches are, however:

1. Microfocus X-ray inspection: see Section 10.10.
2. Microscopy imaging inspection, where the reverse image of the nominal pattern is back-projected by optical mixing onto the substrate for comparison; this technique only applies to low-volume and low-resolution inspection on small substrates.
3. High-resolution wide-area CCD imaging, with at least 2000 × 2000 pixels of 8 bits, combined with a step-and-repeat device for scan-and-inspect operations; frontlighting and backlighting must be combined to differentiate the inks and their colors and obtain circuit pattern discrimination; design rule verification is then applied (see Chapter 9).

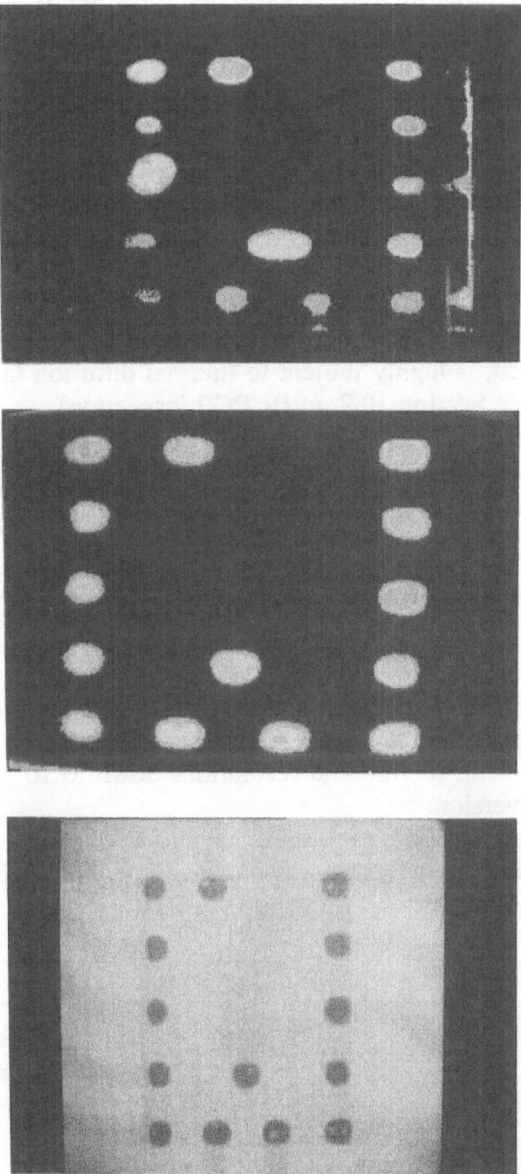

Figure 91. X-ray microfocus image of a flip chip with 14 solder joints represented by spherical blocks underneath the components: white areas within each solder joint represent porosity (top image); gray-level morphomathematical erosion to check on solder pad shape (middle image); bad chip in terms of solder pad shapes (lower image). (Courtesy Nicolet Instrument Corporation.)

4. Substrate integrity test, where the substrate is immersed in a closed-loop fluid system containing a solution in a tank equipped with an anode. A probe is positioned on a pad located on the multilayer and then energized with a pulsating negative bias. All conductor traces electrically connected to the pad identify themselves through a pulsating electrochemical change in the solution that produces a reversible color change. This color image is compared with the one of a good board simultaneously under test. A short exists if more pixels are visible than on the reference, and an open exists otherwise.[85]
5. Infrared thermal imaging, which, although it can sometimes be considered, is highly subject to thermal diffusion in the substrate layers (see Section 10.9 on IR PCB inspection).
6. EBIC generated by SEM in thick films, although the secondary emission is heavily dependent on surface film irregularities (see Section 2.5).

10.11. MICROFOCUS X-RAY INSPECTION OF MULTILAYER PCBs

Multilayer PCBs must be inspected in order to align the layers (e.g., 20) and to center the drilled holes. X-ray-sensitive vidicon cameras give a magnification of up to ×80 and resolutions down to 10 μm, even after fluoroscopic conversion.[86]

The obtained image is processed, first to detect blobs and then to label and position them (Algorithm Label-n), while satisfying simple alignment and design verification rules (see Figures 91 and 92).

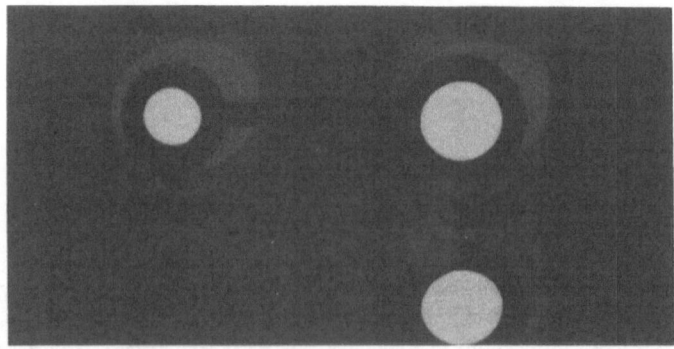

Figure 92. Misregistration in a 24" × 30" multilayer substrate, detected by microfocus radiography, with a 200-pixel line resolution and ×2–50 magnification.

Chapter 11

Inspection for Assembly Tasks

This chapter will review a number of highly useful applications of vision systems to a diversity of electronics assembly tasks.

11.1. MARK READING

Mark reading is a special application of optical character reading (OCR) to the following tasks:

- Wafer mark reading and inspection (Figures 93 and 94).
- Component serial number reading (Figure 95).
- IC package identification reading (Figures 96 and 97).
- Fiducial reading.

Apart from manufacturer logos, most characters, marks, and bar codes have standardized fonts (e.g., SEMI code). The mark reading speeds are, in general, in excess of 100 alphanumeric symbols or bars per second, with error rates of about 10^{-6}, where most errors are due to misalignment, low image contrast, or deterioration of the marks.

Such systems essentially carry out image binarization (Algorithm Thresh-n), followed by longitudinal mark image alignment, framing of an area of interest (AOI) around each symbol, and symbol matching.[87]

Figure 93. Character analysis and identification of a wafer. (Courtesy Cognex Corporation.)

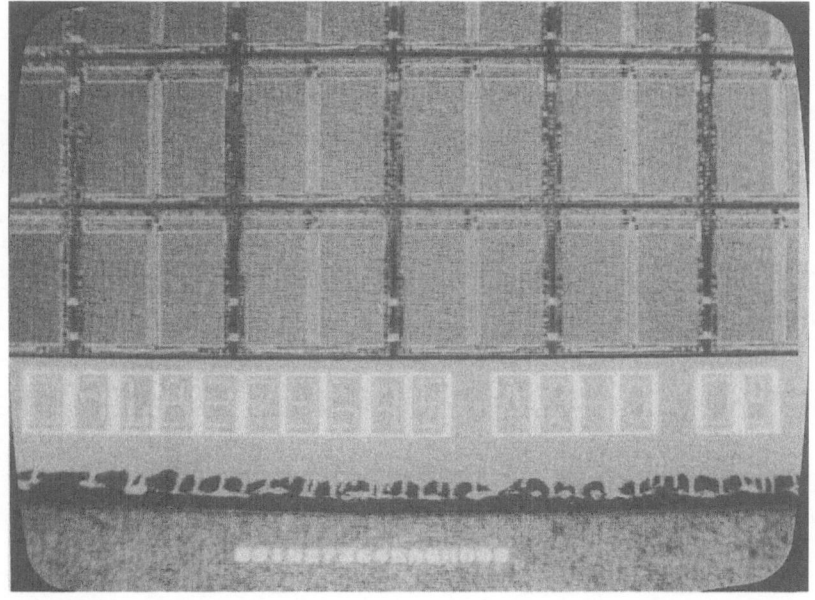

Figure 94.   Closer-look character identification on a wafer. (Courtesy IVS/Analog Devices Corporation.)

Figure 95. Character reading on a component. (Courtesy IVS/Analog Devices Corporation.)

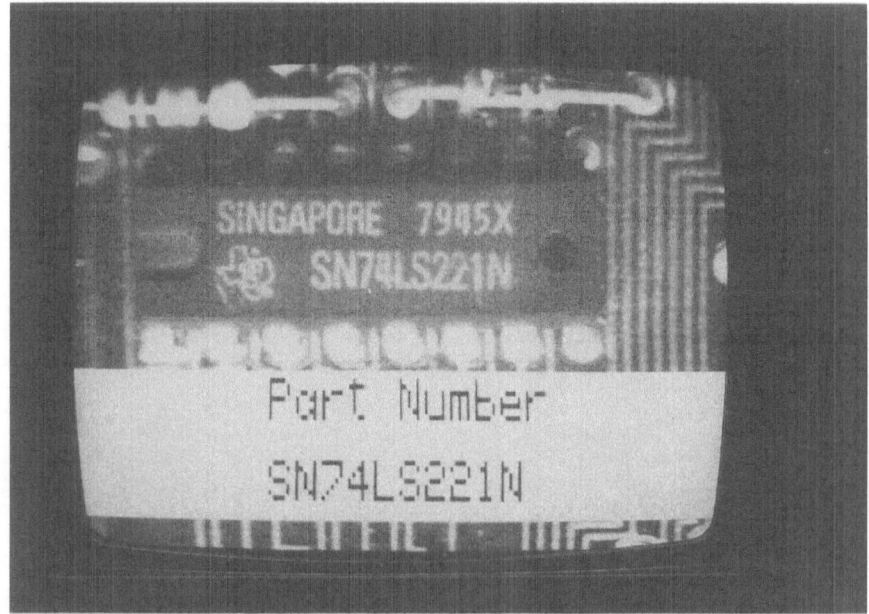

Figure 96. Part number reading. (Courtesy Cognex Corparation.)

Figure 97. Package identification. (Courtesy IVS/Analog Devices Corporation.)

Table 33.   Basic Steps in Label Inspection

Sensor: Video camera.
Attitude: Top-down view.
Illumination: Ring light around camera lens.

1. Thresh-2: threshold selection and binarization.
2. Reg-2: sequential similarity detection, initialized at nominal letter position in reference file.
3. Subtr-1: subtraction.
4. Shrink-1: shrinking.
5. Geom-1: calculation of deformation.
6. Acceptance criteria (Chapter 20).

11.2. 2-D PACKAGE INSPECTION (LEAD, LABEL, PACKAGE MATERIAL)

In the 2-D package inspection step, the component packages are inspected to verify mechanical package integrity (no flaws, delaminations, scratches, bent leads, or knuckled-under leads) and the legibility of the label (also called symbolization). The need for this step arises for passive components, dual-in-line packages (DIP), surface-mounted device (SMD) packages, and most packaging types, especially in connection with insertion, onsertion, or board stuffing machines. The basic steps are given in Table 33.

The lead inspection requirements are:

- Easy adjustment for varying number of leads.
- Measurement of lead tip position accurately to within 1 mil.
- Detection of missing and extra leads.
- Detection of contamination of the leads (ink, paint, grease).
- Detection of solder blobs and solder chips on the leads.
- Discrimination between reworkable and nonreworkable leads.

The label symbol inspection requirements are:

- Recognition of serial numbers.
- Inspection of letter printed in any character font.
- Inspection of arbitrary "logo" symbols.
- Adjustment to different printing processes (stamp, roll, soft touch, laser print).
- Sensitivity to legibility variations.
- Insensitivity to label position, stamping angle, ink color, workpiece position.

The package material inspection requirements are:

- Detection of material flaws (spots, cracks) (e.g., by scanning acoustic microscopy).[88]
- Detection of material delamination (by high-penetration neutron radiography[89] for absorption changes or by ultrasonic microscopy[90,91] (Figure 98).

Occasionally, component matching appears as a specific requirement, e.g., for matched pairs of potted transistors; each component is X-ray imaged in two perpendicular planes, while both in each pair are affixed to opposite sides of a plastic block, which is flipped over between the two views.

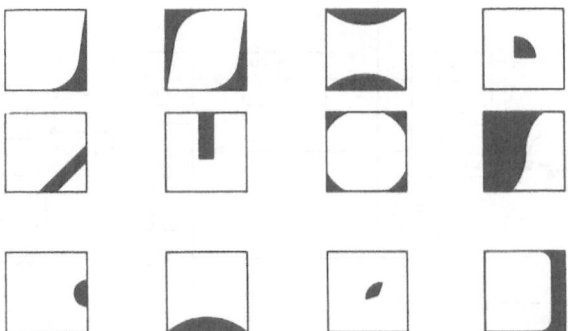

Figure 98. Rejection (top two rows) and acceptance (bottom row) criteria based on the shape of defects found in ceramic packages by ultrasonic reflective scattering. (Note that if the accepted packages have four or more of these defects, as seen in the bottom row, then they are rejected.)

11.3. 3-D PACKAGE INSPECTION BY RANGE/INTENSITY IMAGES

In many inspection problems especially related to the packaging, the image intensity information is not sufficient, either because not all the pertinent information is in the 2-D plane and/or because the 2-D intensity image is corrupted by illumination effects (see also Section 1.5.1). Such problems include:

- Position of component and solder pads.
- Volume of solder paste at solder blobs or solder pads.
- Position of bond wires.
- Substrate thickness and camber/flatness.
- Pattern thickness.

The approach is to improve image segmentation by sensor fusion combining (see Section 5.1) 3-D range images provided by a scanning laser or an array of photoemitting diodes, and 2-D gray-scale images, in order to aim for 3-D feature extraction while achieving insensitivity to color and reflectivity (Figure 99).

However, to resolve, for example, heights of 0.001", a time delay resolution of better than 0.1 ps is necessary for time-of-flight laser ranging. This is presently infeasible (Figure 100). The alternative 3-D

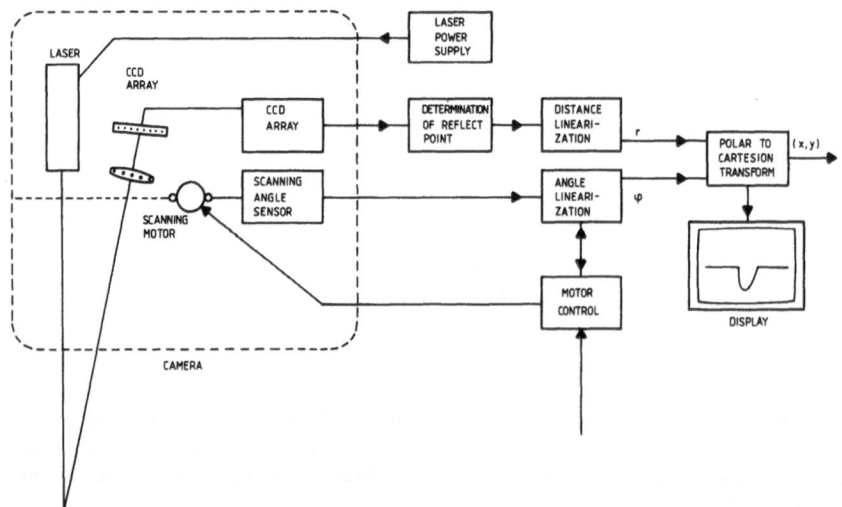

Figure 99. 3-D laser-based imaging and inspection, showing the scan corrections to be performed.

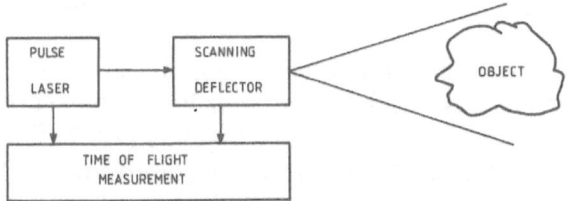

Figure 100. Range measurement by laser time-of-flight measurement.

range determination methods are, therefore:

1. 2-D imaging of lines created by a scanning illuminating laser beam, followed by determination of the height profile in the scene from the displacement of the line of laser light[92] (Figure 101).
2. 2-D imaging when the laser scanner has a synchronously scanned position-sensitive detector (Figure 99); a rotating polygon mirror simultaneously illuminates the area of interest at an angle perpendicular to the reference surface and deflects the field of view of the 2-D sensor along the same optical path, at an offset angle for triangulation. The 2-D sensor collects and measures the offset of the focused spot due to the object height; a resolution of 250 height levels of the total height range can be achieved.[93] The resulting 3-D range images should not be processed like other 2-D images, as 3-D data have extreme slopes in the x–y–z coordinate system. The better coordinate system for range data is the cylindrical polar coordinate system ρ–ϕ–θ, in which level slicing and edge detection still apply.[94]

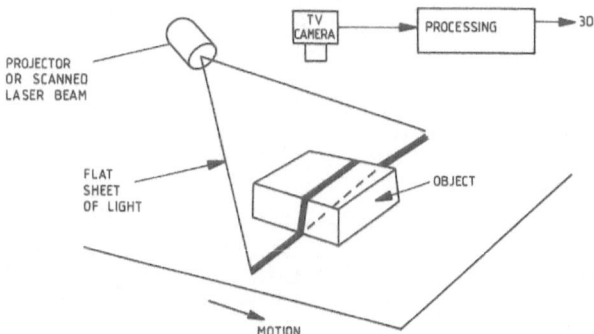

Figure 101. Structured light 3-D imaging by analysis of the nonconnected light stripes projected onto the object.

11.4. DIE ATTACH, SOLDER, AND BONDING INSPECTION

Another type of inspection is the testing of the bonding, solder, or other attach between the components, their leads, the substrate or die, and the carriers, such as tape automated bonding (TAB).[95] The areas of interest are the bonding, solder, or die attach patterns themselves, in that we do not treat here the defects in the bonds themselves, but the defects related to disbonds. A disbond can be thought of as a pancake-shaped void, the thickness of which is less important than its area. A disbond impairs not only the electrical connection, but also the substrate's ability to act as a heat sink for the component. Actual physical separation of the component from the substrate is a less common consequence; changed electrical current and overheating caused by a disbond or thermo-compression will eventually change the inputs to the component and cause it to fail.

Typical detection approaches involve:

- Scanning laser acoustic microscopy (SLAM), with 25-μm resolution at 100 MHz, which reveals disbonds or lead bonding problems as strong reflections with very dark spots in the AOIs[5,88] (Figures 102–105).

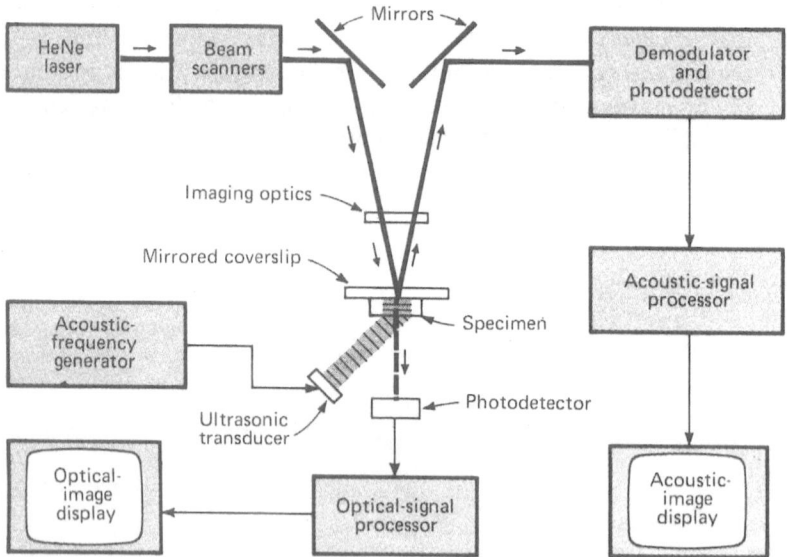

Figure 102. Scanning laser acoustic microscopy (SLAM) for disbond inspection. (Courtesy Sonoscan.)

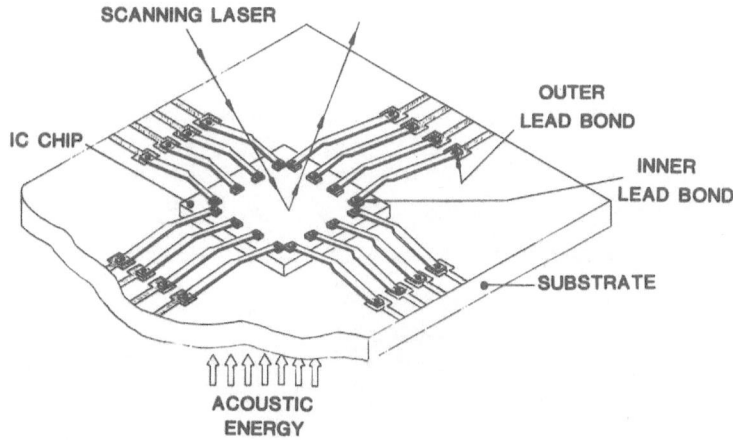

Figure 103. Chip bonded to substrate by TAB, and LSM or SLAM imaging thereof.[5] (Courtesy Sonoscan.)

- X-ray radiography, which can also detect defects internal to the bond/solder itself, besides detecting missing/misaligned/bent components and leads.
- Reflected polarized IR light, using the IR transparency of silicium to observe the bonded region[22] (Figure 106).
- IR thermal response of solder or bonding volumes irradiated by a long-pulse-duration directed laser beam (30–180 ms) (Figures 107 and 108). The solder integrity and voids are characterized by the joint's peak temperature at irradiation and the rate at which it cools.[96,97] Physical templates are used: if a joint contains insufficient solder, its temperature will rise higher than will that of a good joint; if there is excessive solder, the joint's temperature will not not rise sufficiently. The inspection speed is about 10–15 joints per second, and optical filtering is required to make the sensor blind to the laser's illumination.[98]
- Color image processing of solder joints.[99]
- Standard video images (Figure 109).

In all cases, area-of-interest processing (Algorithm AOI-1) is applied to the die attach and bond attach areas. Also, morphomathematical algorithms (Algorithms Morph-n, Skeleton-1) must always be applied for the removal of border effects and to characterize shape.

The basic steps for lead inspection are given in Table 34. The basic steps for wires and dies can be found in Table 35.

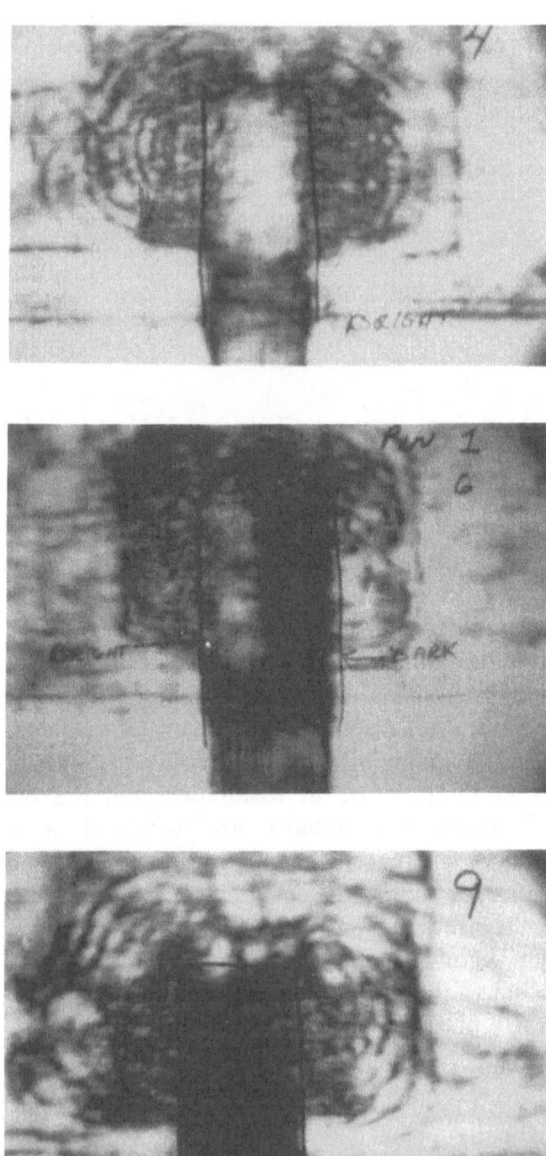

Figure 104. Scanning laser acoustic imaging (SLAM) of lead solder reflow bonded to an alumina substrate[5]: good lead (top), marginal bond (middle), bad bond (bottom).

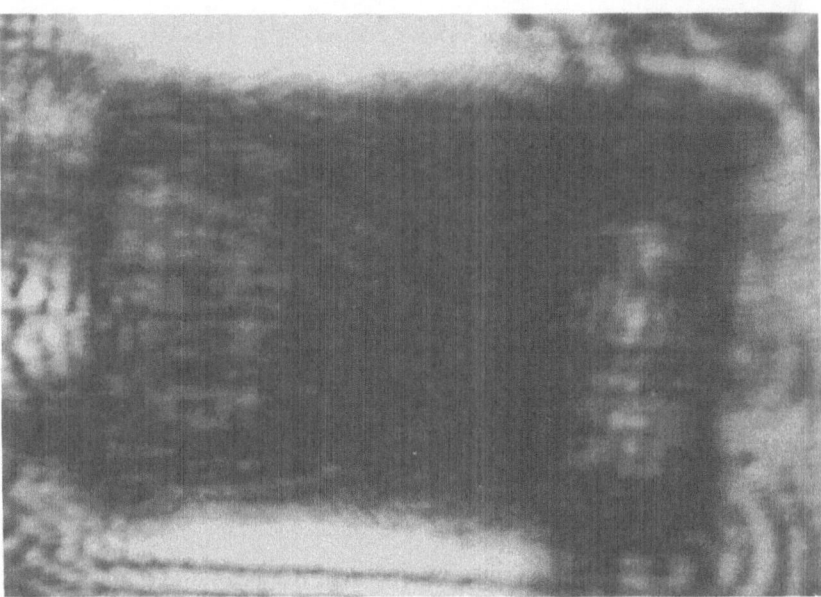

Figure 105. Acoustic micrograph showing the bonding of a chip capacitor to an alumina substrate: top image locates the capacitor; bottom image shows a bright patch at each end which outlines the contact areas between the capacitor and the substrate.

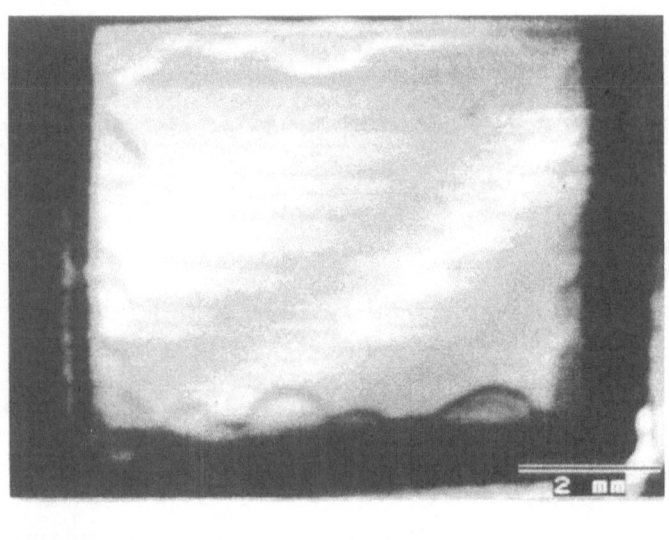

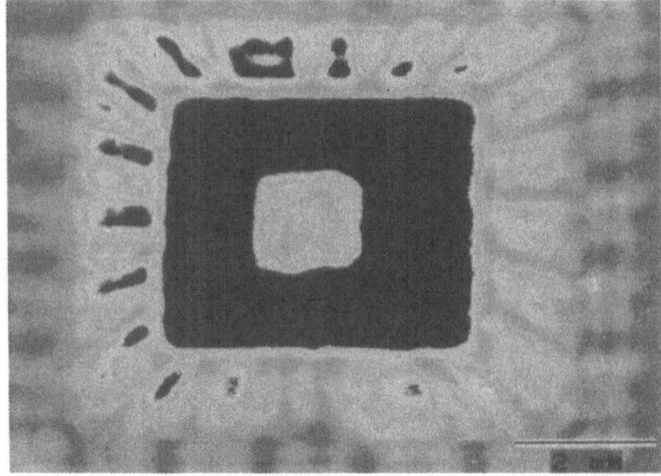

Figure 106. Acoustic images of a leadless chip carrier: Top image through the lid shows uneven bonding around the edges; bottom image through the base (pseudo-color coded) shows partial bonding and uneven lead adhesion. (A color version of this figure can be found at the end of Part I, after Chapter 12, facing page 205.)

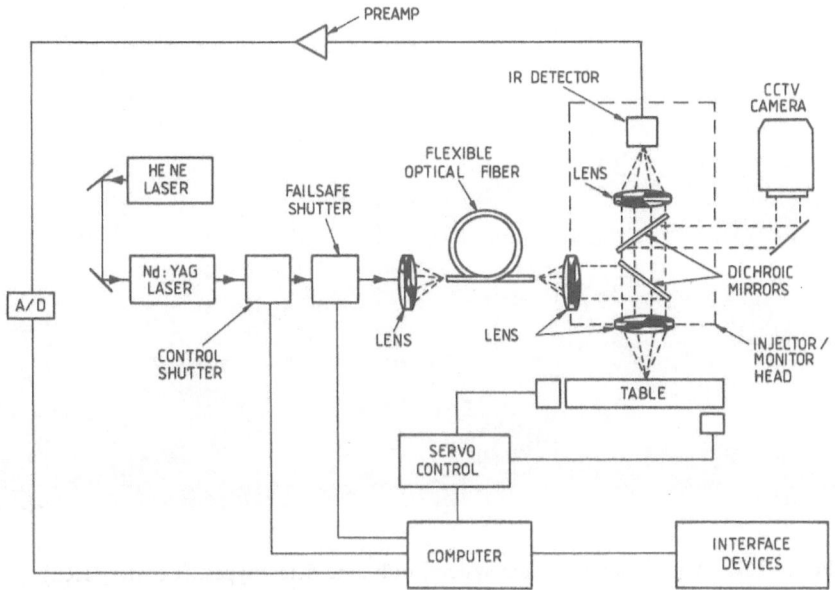

Figure 107. Block diagram of Vanzetti Corporation solder laser/inspect system.[96] (Courtesy Vanzetti Corporation.)

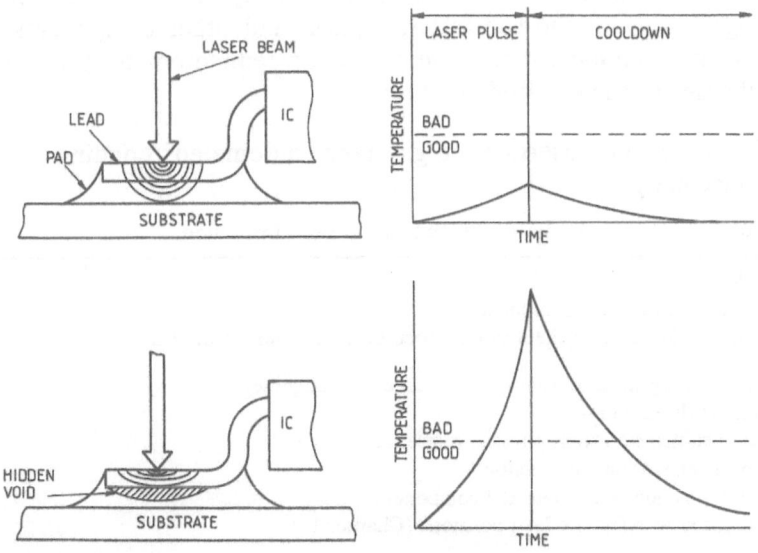

Figure 108. Thermal time signatures of a normal (top) and a defective (bottom) joint obtained with system in Figure 107.[96] (Courtesy Vanzetti Corporation.)

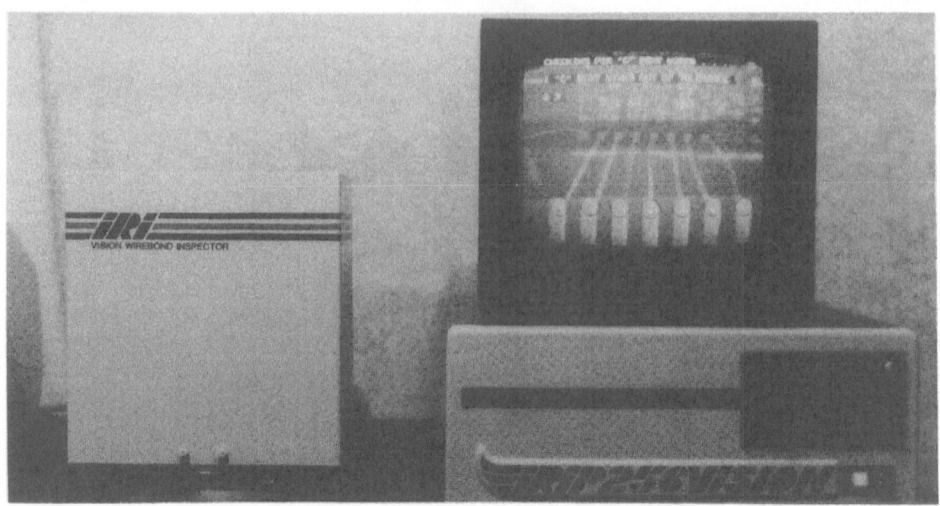

Figure 109. Wire-bond inspection system, with nine defect types. (Courtesy IRI.)

11.5. COMPONENT PLACEMENT

11.5.1. Applications

Component placement aids, consisting mainly in package and lead or tool alignment with the substrate(s), pads, and other components, are required in a number of fully automatic or semiautomatic procedures. Typical cases of such procedures are:

- Component insertion, e.g., tape automated bonding (TAB) machines.

Table 34. Basic Steps in Lead Inspection

Sensor: Video.
Attitude: Bottom view of components.
Illumination: Illuminating sheet of light focused at the ends of the leads.

1. Mechanical registration from package edges and alignment.
2. Thresh-*n*: thresholding.
3. Subtr-1: subtraction from physical reference.
4. Morph-1: expansion–contraction.
5. Label-2: position of a connected component.
6. Comparison to reference lead positions (Chapter 15).

Special aspects:
- Steps 4 and 5 are executed from the middle of the image and work outwards.

Table 35.   Basic Steps in Wire and Die Bonding

Sensor: Top-down or lateral video sensor.
Attitude: $X-Y$ servomotor-driven table (see Section 6.1).
Illumination: Autofocus on bonds or die surface.

0.1 Thresh-n: storage and thresholding of reference grid of nominal bond and wire positions, e.g., 128 × 128 array.
0.2 AOI-1: masking out of insignificant repetitive features.
1.   Thresh-n: acquisition and thresholding of live bond and wire image, for same sensor position.
2.   Reg-1: correlation of step 1 and step 0, and location of points of highest correlation, with rotation invariance.

Special aspects:
• Possibility for LUT-based enhancement in steps 0 and 1.
• Selection of various pixel resolutions and AOIs under system control, because of changing device and pad sizes.
• Multiplexed access to correlation step 2.

- Component onsertion, e.g., in the case of surface-mounted devices (SMDs), small-outline transistors (SOT), small-outline ICs (SOIC), plastic leaded chip carriers (PLCC).
- Paste, pads, e.g., for SMDs.
- Computer-aided wafer or board repairs.
- Wire bonders and wire changes (thermosonic or other types).
- Die bonders.
- Die alignment on wire bonders.
- Desoldering.
- Probers of automatic test equipment (ATE).
- PCB drilling with hole placement feedback.
- Pick and place systems.
- Connector mounters.
- Laser trimmers (see Section 4.11).
- Wafer saw alignment.

Examples are given in Figures 110 and 111. The basic steps in automatic component placement are listed in Table 36.

In automatic component placement, the visual feedback to handling and positioning is unattended; in the semiautomatic cases, guidance waypoints are projected onto the package and/or substrate in graphic overlay telling the operator how to adjust the positioning. Translation and rotation offset are calculated and sent to the placement system on the basis of image features or visually acquired generator signals.

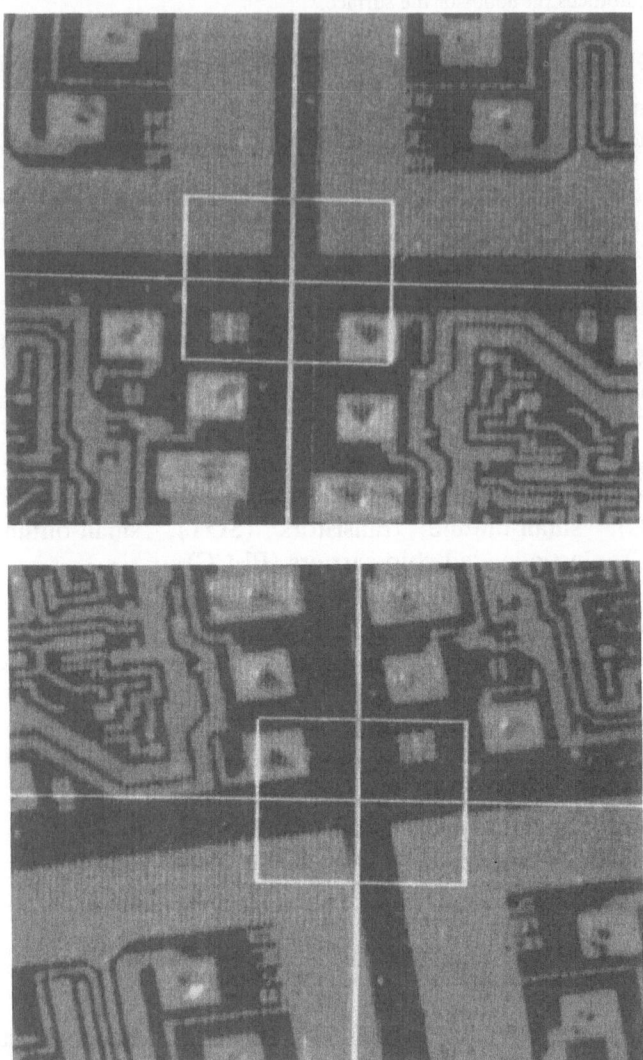

Figure 110. ASIC alignment, before and after offset calculation (Algorithm Offset-1).

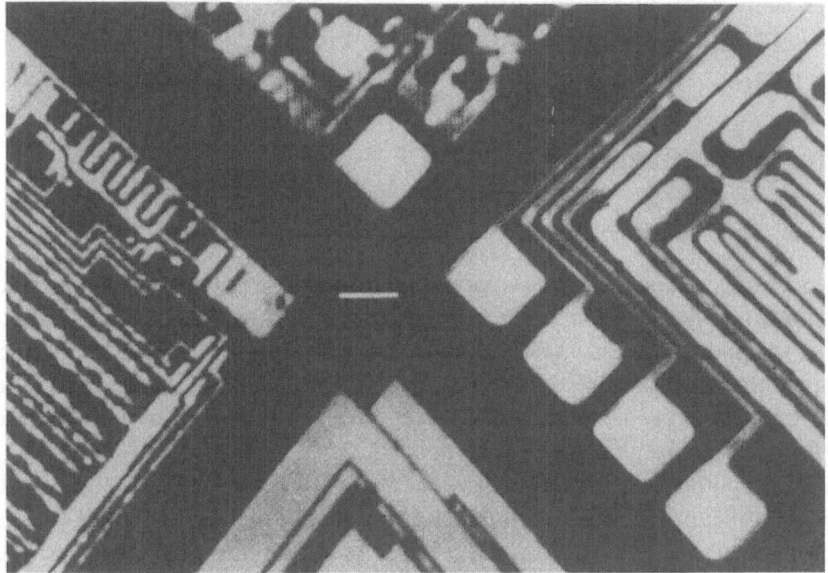

Figure 111. Digital alignment figure, with alignment on point between patterns.

Table 36. Basic Steps in Automatic Component Placement

Sensor: High-resolution CCD or laser scan or PMT.
Attitude: Top-down view with large depth focus range.
Illumination: Narrow beam, highly homogeneous from top or back.

0.1 AOI-1: AOI creation from component floor plan.
0.2 Autofocus in step 0.1.
1. Reg-3 multiresolution resolution alignment in step 0.1.
2. Edge-1: filtering-based edge extraction.
3. Offset-1: offset calculation.
4. Edge-3: subpixel edge extraction.
5. Repeat step 3 until positioning.

Special aspects:
• Step 4 can be replaced by the correlation-based Reg-1 and by finding the peak in the correlation image by thresholding the correlation gradient image.
• Rotational error is minimized by selecting the pattern in step 3 such that the intersection points are far apart.

Component alignment can also be performed by alignment with backlighting, e.g., IR light.

11.5.2. SMD Placement

As mentioned above, SMD placement (Figures 112 and 113) requires vision because these packages, without legs to guide them into position, can float sideways on their adhesive as they are glued in place.

The measurement resolution required is about 0.005″, with a measurement repeatability of ±0.0010″ (at ±3σ); at the same time, the field of view must be wide enough to cover the area over which these performances apply (e.g., 1″ × 1″ or larger).[1]

The spatial relationship between the leads and the SMD body can vary significantly due to deformation of the leads; therefore, the component body cannot be used reliably to define the positioning of the electrical contacts with respect to the pads on the circuit board. Furthermore, the view of these pads is partially blocked by solder paste and by the component itself. The component position and orientation must therefore instead be related to visual reference marks, detected by vision, which are part of the same circuit board. These visual reference marks are called "fiducials." This compensates for variations between the artwork and the substrate, and variations between the circuit board substrate and the inspection system fixturing. Local fiducials must be present in the sensor field of view.

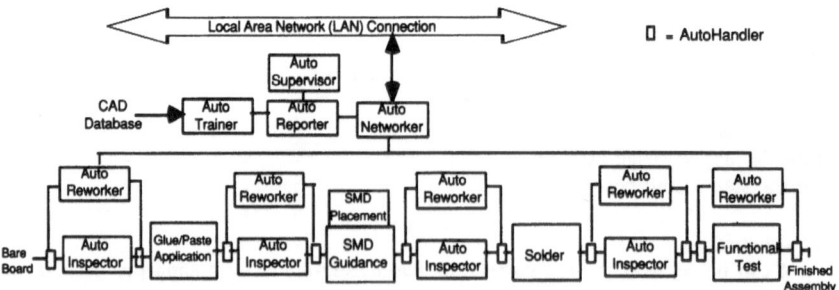

Figure 112. Placement of inspection stages in a surface mount technology (SMT) line. (Courtesy IRI.)

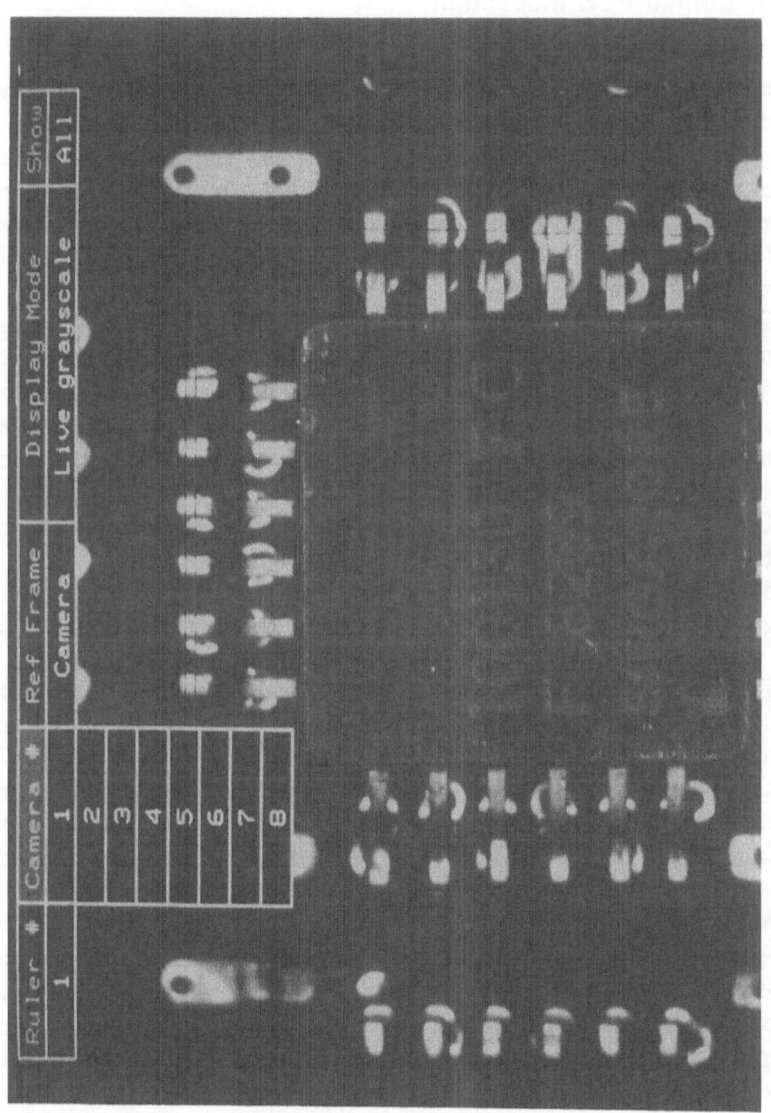

Figure 113. SMD placement verification by three rulers (Algorithm Ruler-1): two on top, one on the side. (Courtesy AdeptVision Inc.)

11.6. MISSING OR IMPROPERLY MOUNTED COMPONENTS OR LEADS

11.6.1. Stuffed PCB Inspection

Many assembly operations require checking for missing or improperly mounted components or leads. The most important feature to detect is the absence of a needed component/lead. In other cases, the component/lead may be mounted right but with the wrong pin orientation or may be skew with respect to its nominal orientation. Examples of operations which require this type of checking capability are control prior to soldering, postverification of errors of an automatic insertion machine, and rework stations.

Common to most systems is the practical facility whereby missing or improperly mounted components are physically designated by a light spot pointing to the nominal position on the board. A message is given about the type of defect. If a component is missing or must be replaced, a parts bin presenter automatically presents the correct one to the machine or operator.

Optical character reading options to identify component numbers may tell if the mounted components are the right ones.

The implementations must be made independent from the board background reflectance, from the component package reflectances, and from variations in the component height above the substrate.

The learning must be from a stuffed board layout CAD file with labels at each component/lead location giving the package outline size, orientation, type, and height after insertion.

The solution meeting these requirements involves area of interest processing (Algorithm AOI-1), with the AOI boundaries generated from the CAD layout file, and special lighting, such as presence-from-shade or X-ray imaging, with knowledge-based accept/reject image analysis (Algorithm Rule-1).

Among the lighting possibilities, the most robust is the presence-from-shade approach, which involves checking the presence and orientation of components, not from the direct image itself, but from the shade they project onto the substrate or other surfaces, from known illumination directions (achieved by, for example, laser, X-ray, or collimated beams). The dark shade is much easier to detect robustly than component houses with varying reflectances. If needed, the lighting can be multiplexed among different directions. The nominal shade outlines can

Table 37.   Basic Steps in Detection of Missing or Mismounted Components

Sensor: High-gain CCD camera.
Attitude: Top-down view.
Illumination: Presence-from-shade, through oblique focused lighting from one or several
     directional sources.

0.1 AOI-1: AOI segmentation from the CAD layout.
0.2 Thresh-2: thresholding within each AOI, for the reference board, to find the shade.
1.   Thresh-1: thresholding within each AOI, with the threshold found in step 0.2, to find
     the shade.
2.   Subtr-1: comparison of the binary shade images with step 0.2.
3.   Shade-1: calculation of the shade features in step 2.
4.   Rule-1 or Class-1: classification of the defects from step 3.
5.   Pseudo-1: pseudo-color coding of the missing or mismounted component footprints in
     step 0.1.
6.   Activation of optical or mechanical designation of the missing or mismounted
     component, by pointing at it.

easily be calculated by ray tracing on the basis of the 3-D component outline, the component file, and the light source locations.

The steps for detection of missing or mismounted components with the use of presence-from-shade illumination are listed in Table 37.

11.6.2. Other Applications

11.6.2.1. Underside of Stuffed PCB

One case is the often forgotten inspection of the underside of stuffed PC boards to check that leads have correctly penetrated the board to the tracks on the far side and that they are securely turned over to the right side.

11.6.2.2. Displays, Keyboards, and Connectors

The test of displays, keyboards (Figure 114), and connectors (Figure 115) is a special case of the above, which however also includes keyswitch character or tone actuation; in other words, vision is combined with touch and electrical simulation.

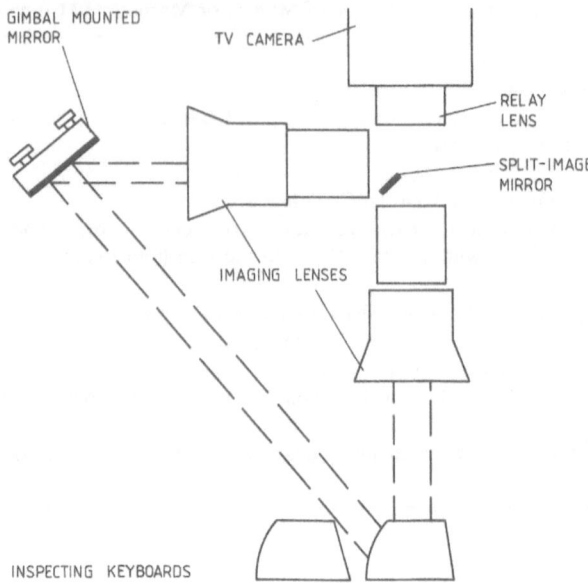

Figure 114. Keyboard inspection system; the keys are seen in profile at the lower right.

11.6.2.3. Wire Bond Inspection

Another case is testing of hybrid circuits, DIPs, and PLCCs during or after wire bonding. Three-dimensional features are obtained by focused directional lighting and shading. The task is simplified by the linear structure of the AOIs. Defects detected include missing wire, multiple wires, "C" bent wires, broken wires, high loops, and misregistered bonds.

Figure 115. Connector inspection system in operation; the stack is seen on the left side before transfer; a connector image appears on the display. (Courtesy C. E. Johansson AB.)

Chapter 12

Knowledge-Based Printed Circuit Board Manufacturing

In this chapter, the printed circuit board manufacturing tasks (Figure 116) where artificial intelligence (AI) could be or is applied are examined. In order for such tasks to be considered suited to the application of knowledge-based techniques (Chapter 8), the following conditions should be met:

1. The tasks should already be computer controlled or computer aided to some extent.
2. The payoff expectations should be high because of resulting savings and/or quality improvements through qualitative knowledge.

The sequence of operations is depicted in Figure 117.

12.1. PCB DESIGN

12.1.1. PCB Design Verification and Routing

The following tasks are normally implemented in PCB CAD systems, with shop floor editing possibilities:

1. Routing the PCB interconnections	*Framework:* Automatic routing should only require as inputs the boards' physical characteristics (dimensions, spacing between traces, number of signal layers,

187

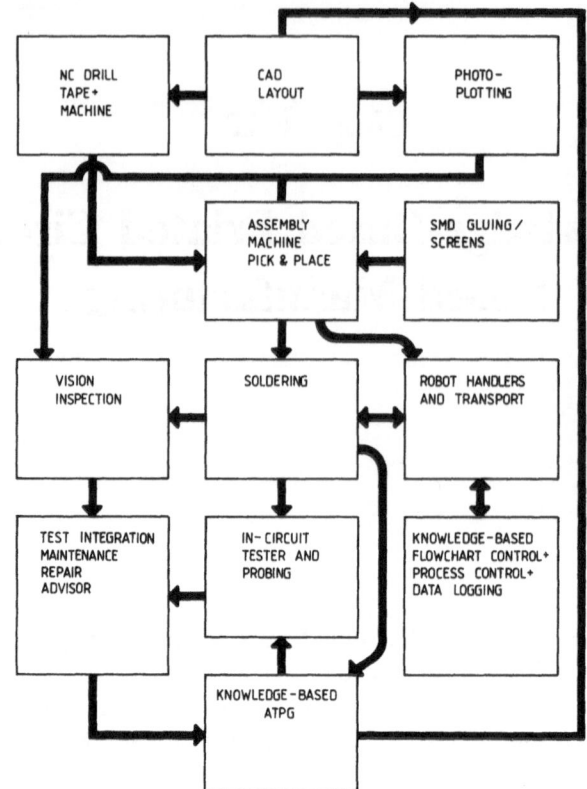

Figure 116. Overall flow of knowledge-based PCB manufacturing.

and CAD-generated board layout). The goal is to maximize the completion rate on the first prototype boards.

Knowledge bases:

(a) Routing is selected through the adjustment of approximately 30 costing factors in the product's maze and line probe algorithms. The router sets such casting factors as: grid size; via grid, which reserves specific grid locations for placement of vias; and pin keep-away, a procedure that leaves routing channels open in order to facilitate the automatic placement of etch and future engineering change orders.

(b) Design preplanning is also necessary, so that the final product offers the same patterning and aesthetic quality as found on designs from skilled users. One way this is accomplished is through the independent

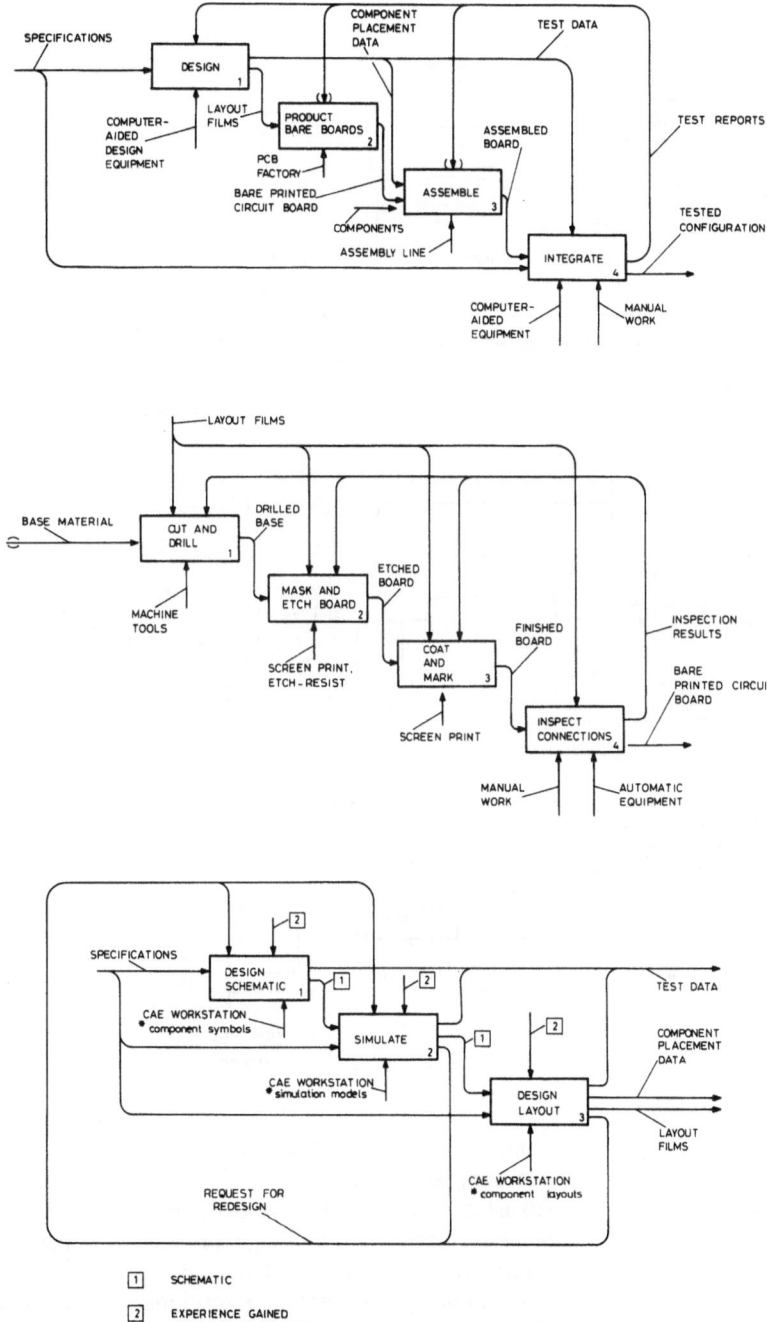

Figure 117. Detailed flow of knowledge-based PCB manufacturing, in standard SADT (Specification Analysis Design Tool) flow diagram notation; figure is in six parts.

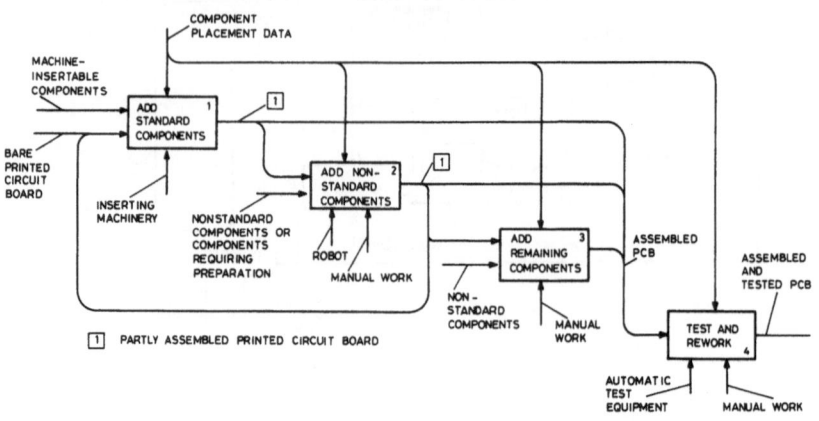

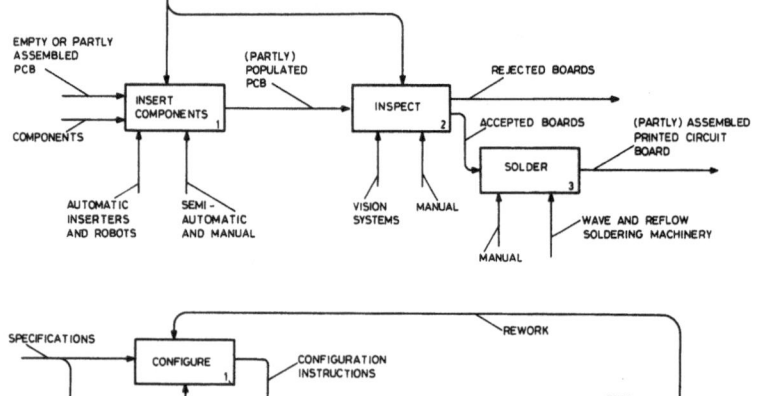

handling of horizontal and vertical connections without using any vias.

(c) Blockages must be avoided, by designating legal via sites early in the design process; this preserves board real estate and accelerates design.

Programming: Routing algorithms in compiled Prolog; user interface in Pascal.

Inference: Forward chaining.

2. Design verification

Framework: The PCB designer must be able to analyze and verify the correctness, timing, speed, etc., of PCB designs. CAD workstations have analyzing tools, but the designer has to verify the design.

Feasibility: A knowledge-based system could use the analyzing tools of a CAD workstation to verify the PCB design characteristics and suggest improvements. Building such a system is difficult because of the complexity of PCB design and because of the limited interfacing capabilities of CAD workstations.

3. Design for testability

Framework: It is essential that PCB designs, both manual and CAD based, be made more testable. This can be carried out in an interactive, knowledge-based fashion.

Knowledge bases: A text base gives design and test procedure guidelines, accessible through queries, for LSIs, SMDs, and analog components as well as mechanical components.

Inference: Simple keyword-based or relational queries.

12.1.2. PCB Layout

1. SMT CAD
   adjustments

Framework: Conventional PCB CAD systems cannot, in general, do the layout for components mounted on both sides of the board. Therefore, packaging- and assembly-related knowledge must be accounted for. Also, surface mount technology (SMT) boards have off-grid data, which makes CAD adaptation difficult.

Knowledge bases:
- Library of component outlines; SMDs, however, are not standardized.
- Set of footprints taking into account the tolerances of placement machines.
- Pad size adjustments to soldering process (wave soldering requires larger pads than reflow).

Feasibility: The knowledge base corrections are difficult to implement on top of existing SMT CAD systems.

Inference: Mixed chaining.

2. Adjustments for analog circuits

Framework: Conventional PCB CAD systems have originally been developed for digital PCB layout design and cannot always cope with the rules of placing analog circuits. When using analog circuits, functional circuit groups must be placed near to one another. Analog circuits often contain many discrete components which tend to be irregularly shaped and sized, which makes the placement more difficult.

Knowledge bases:
• Library of component sizes and shapes.
• Design rules, e.g., concerning functional circuit groups and safety regulations (analog PCBs often have high voltages).

Programming: A high-level tool with good graphics is required. A frame-based system is preferred, because the component family characteristics and differences can be implemented using inheritance. Difficult to implement on an existing CAD tool.

Feasibility: The design rules, e.g., groupings, are easy to implement, but automatic placement is difficult.

12.2. BARE-BOARD PCB PRODUCTION

12.2.1. Tooling

1. Bare-board drilling

Framework: Numerically controlled machine (CNC) tapes must be generated for board drilling, interfaced to CAD, phototooling, and possibly automatic bare-board inspection.

Programming: Prolog and conventional programming.

Feasibility: Commercial products are available.

Inference: Constraint propagation.

2. Tool-hole locators

Framework: Tool-hole location systems (ATHL) use robotics and vision to align tooling holes in multiple-layer PCBs.

Feasibility: 4:1 productivity gains in labor hours are reported.

Inference: Constraint propagation.

12.2.2. PCB Inspection

1. Bare-board inspection

Framework: Bare boards must be checked for continuity and other defects, against either a physical reference board or a CAD-based layout.

Knowledge bases:
- CAD layout.
- Physical track dimensions.
- Design verification rules, with selectable rejection criteria.

Programming: Conventional, and microprogramming of bit-slice or digital signal processors (DSP).

Feasibility: Commercially available, including for multilayer boards although speeds, performances, and costs vary a lot; the algorithms essentially rely on morphomathematical image processing operations; the sensors are either laser or video.

Inference: Backward chaining.

12.2.3. Parts and Materials Transport

1. Deliver components and materials to work cells

Framework: In a decentralized assembly and test repair plant organization, work cells need to be supplied with components and materials, while work cell inventories and other plantwide buffer stocks are minimized.

Programming: OPS-5 or LISP, mixed with conventional programming.

Feasibility: Bar-coded boxes are automatically selected and routed through each appropriate work cell by means of a bidirectional conveyor loop, with adjacent peripheral work cells. Commercial products are available.

Inference: Backward chaining.

12.2.4. PCB/PWB Assembly Operations

1. Operator override in case of automatic inserter errors

Framework: In case of inserter errors, detected on-line or after subsequent inspection, the assembly line operator may end up facing a situation with cognitive overload; this arises when the rate of alarm messages in critical operations goes beyond the operator's capability of interpretation or of recalling timing/sequence information. Operator override also leads to optimization of the assembly operations with respect to available operating resources.

Programming: Some researchers in industry address this problem by providing an intelligent interface between the alarm/situation assessment expert system and the process control equipment. The scheduling of prioritized tests, checks, and tasks is done in the expert system; the real-time requirements are off-loaded from the expert system to the user interface.

Feasibility: Difficult, because of real-time requirements and need for tracing data. Current LISP workstations cannot easily deal with interrupts from the outside, in the way process control computers do. LISP garbage collection over long periods of time is also a problem.

Inference: Research is on time-dependent inference.

2. Assembly of printed wire boards (PWB)

Framework: The goal is to produce on-line assembly instructions from a CAD data base. Secondary results should be bills of materials, engineering notes, and side-view explosions of the printed wiring assemblies. The expert system should approve a design as manufacturable and download the task to the production floor or return the design to engineering for changes.

Feasibility: Manually creating a PWB assembly plan takes three weeks, with later changes increasing the time required. The expert system can do the job consistently, without errors, in one week.

3. PWB placement and routing process

Framework: Automatic placement techniques tend to minimize wire length and wire density in multilayer boards.

Programming: LISP or conventional programming or LISP first for prototyping.

Feasibility:

(a) For placement, interactive techniques are used, especially in relation to the min-cut algorithm.

(b) For PWB routing, in multilayer boards, the usual approach involves fixed vias and orthogonal routing. The most popular routing process is line search, followed by maze running and channel routers.

Inference: Mixed chaining.

12.3. INTELLIGENT WAVE SOLDERING

The introduction, in 1984, of computer-controlled wave soldering machines led to an assessment of which process tasks would be suited to the application of knowledge-based techniques in terms of the gains to be achieved in precision, repeatability, diversity, and setup time.

1. PCB tracking

Framework: PCB tracking facilitates counting of PCBs, intermittent fluxer operation, actuation of fluxer air knife, and temperature sensing on the surface of the PCB for closed-loop preheat control. Other functions are intermittent solder wave operation, verification of readiness for a new process setup, and process simulation (off-line). Heat-resistant, photoelectric, or capacitive sensors coupled with time delay relays are required at each board recognition point along the conveyor.

Programming: Conventional programming sufficient, with alternatives in object-oriented programming.

Feasibility: Feasible by classical instrumentation, but infinite tracking of a single PCB and conveyor traffic monitoring are possible (with bar code reading) only by computation of PCB length and position. Unique PCB shapes, cutouts, and positions on carriers can be accounted for via dedicated software.

2. Conveyor speed

Framework: Closed-loop control is achieved through the use of a computer-interfaced pulse counter. The actual speed (rpm) is compared to the set point speed, and corrective action is implemented automatically for each process run.

Knowledge bases: Corrective actions, updated from layout characteristics, written in rule- or frame-based form.

Feasibility: Feasible by a combination of simple feedback controller software, a set of state identifiers, and by rule-based corrections.

Inference: Forward chaining.

3. Conveyor width

Framework: Conveyor is automatically set for each process run. The conveyor position is sensed by a rotation pulse counter and referenced to a zeroing switch. The setup accuracy is determined by the product of the number of previous setups and the possible error in each setup. A simple setup's error equals the linear distance traveled in one drive revolution divided by the number of sensor pulses per revolution. When the error reaches a predetermined value, the conveyor position is reset via the zeroing switch before proceeding.

Programming: Conventional, or PLC control language.

4. Fluxer controls

Framework: Control functions associated with the fluxer module include temperature-compensated specific gravity control, fluid level control, foam, wave height, and air pressure monitoring, and self-cleaning cycle.

Knowledge bases: State filters, compensation actions, alarms, logic commands in proper sequence. Predicate form seems adequate.

Programming: Symbolic language, especially logic programming.

Feasibility: Specific gravity (SG) monitoring requires density and temperature sensors with ADCs. Set points are recalled with each process change. SG tracking, temperature compensation, warnings, alarms, and messages are all to be software controlled. Detection of empty flux or thinner containers, thinner addition, and automatic cleaning cycle of the fluxer are also software dependent.

Inference: Primarily backtracking.

5. PCB preheat

Framework: Each process consists of dedicated, user-defined thermal parameters which shape the control loop. These parameters include PCB temperature set point, zone-1 starting intensity, zone-2 intensity factor, PCB emissivity adjustment, and PCB temperature sensing area. The preheating intensity is logically coupled with conveyor speed.

The preheat control loop operates as follows: The pyrometer samples the PCB's topside surface temperature at multiple points along a defined area. The samples are filtered to eliminate readings of components, cutouts, and through holes. The remaining samples are weighed to yield the final value, which is normally verified against a contact sensor during the setup procedure. The intermediate temperature is compared to the intermediate set point, and the zone-1 intensity is readjusted to correct for the difference for the following PCBs. Simultaneously, zone-2 is actuated to make up the remaining differential.

Programming: Control hardware or conventional programming, including expert controller firmware.

Feasibility: A double-zoned preheating module separated by an infrared surface sensor station provides true closed-loop temperature control of PCBs. Each preheat zone is made up of longitudinally mounted quartz tubes or lamps housed in an insulated reflective cradle and is covered by a topside insulated reflective tunnel. The temperature sensor station consists of a pyrometer housed in a forced-air-cooling antireflection tunnel, mounted above the conveyor level.

6. Solder temperature

Framework: Temperature readings are provided by a single thermocouple interfaced to the computer via data logging. Set point control, out-of-range indications, alarms, temperature readout, and external interface to the process data base can be software controlled.

Programming: PLC commands or conventional programming.

7. Solder wave height control

Framework: Closed-loop control is achieved through the use of a computer-interfaced rotations pulse counter. The actual rpm reading is compared to the set point rpm, and corrective action is implemented by digital-to-analog direct voltage regulation of the motor. The wave height set point of each process may be entered numerically or read directly from a fine-tuning analog dial. Process wave height is implemented automatically for each run.

Knowledge bases: Listing of set points for the process, in a fact base, manipulated by process and layout/size-related configuration rules.

Programming: Conventional, but with interfaces to intelligent query language.

Inference: Mixed, with exhaustive constrained search.

8. Operator console

Framework: Operator inputs are entered to a standard or process-dedicated keyboard. Outputs are displayed on a standard CRT or dedicated display. Data base format process retrieval provides complete protection against erroneous setup of process parameters; bar code interface permits fully automated process setup without human intervention; software-monitored interlocks increase safety and process integrity; multilevel security codes ensure system level and process level protection against nonauthorized intervention; remote control capability permits the operator increased mobility; communication with other computers provides multistation remote monitoring and control.

Knowledge bases:
(a) Rule-based descriptions of soldering configurations for different component/packaging types used, as well as for different board track parameters.
(b) Unification rules, for trade-offs between configuration requirements.
(c) Explanation facilities.

Programming: Mixed programming in a CLOS-like environment, supplied with setup/configuration verification rules, as well as alarms.

Feasibility: Menu-driven man–machine interface is state of the art, but users have indicated the need for CLOS-like hierarchical process setup and monitoring representations as well as for easier editing.

Inference: Mixed.

9. Flow-soldering diagnostics

Framework: The problem tackled is the diagnostics of faults in flow-soldered PCBs.

Inference: Attempts to use a shallow rule-based approach have proved unsatisfactory.

10. Electroless copper deposition diagnostics for PWB

Framework: Electroless copper deposition (ECD) is a 20-step wet chemical process, required for all printed wiring boards (PWB). Both sides of the boards must be plated flawlessly, but each of hundreds of through holes must receive perfect plating to provide conductors from one side of the board to the other. The wet processes are subject to drifting.

Programming: Expert system shell.

Feasibility: Rule-based approach, where the human operator must first enter the situation on the chemical line, through traces of key deposition variables, labeled as "stable," "erratic," or "cyclic." The operator must also enter gauge reading levels (copper concentrations, bath temperature, sample temperature). The expert system makes diagnosis and suggests corrective actions (cleaning solutions, catalysts, copper solutions, temperatures, and concentrations). On-line sensors and statistical control charts serve to monitor key variables. Readings are accessed through a touch screen.

Inference: Classification procedure.

12.4. STUFFED PCB INSPECTION

1. Inspection of stuffed/populated PCBs

Framework: Machine vision may replace workers in the checking of stuffed/populated boards prior to soldering to ensure that all components are in place and that pins are in sockets. Component orientation can also be checked. Beyond such verifications, vision can also check on dimensions and shapes which must meet prescribed values. The motivation for such inspection is that among all assembly errors, insertion errors account typically for 55%, missing components 20%, wrong polarities 15%, and wrong components 10%.

Knowledge bases:
- Reference patterns or boards.
- Dimensional tolerances.
- Information about illuminations and sensors.

Programming: Assembler language.

Feasibility: Commercial products are available; the positioning speed is about 20–80 components/min within ± 25 μm of designated locations.

Inference: Operates basically by thresholding, registration, and matching/image subtraction.

2. Solder joint inspection

Framework: Solder joint inspection may be required when solder defects are significant.

Programming: Assembler language.

Feasibility: The idea is to sense, in the IR domain, temperature increases of solder joints after lasing. These increases are compared by computer to reference values of a good solder connection of that joint. An adjacent fixture marks any unsatisfactory joint with a blue dot, for repair. (See Section 11.4.)

12.5. TEST AND REWORK

1. Burn-in production   *Framework:* The burn-in process flow involves
   scheduling   mounting the boards in the burn-in ovens, loading the
                             burn-in boards, burn-in over a fixed duration, un-
                             loading, dismounting, and transfer of the boards to
                             final testing. This process requires weekly and
                             monthly planning, due to the time involved in burn-in
                             and to the burn-in capacity involved.[100]
                             Knowledge bases: Scheduling procedures, burn-in
                             system limitations, precedence of operations in burn-
                             area, availability of equipment.
                             Programming: Data base interface and expert system
                             shell.

2. Scheduling and test   *Framework:* A simple test rework work cell consists
   program selection in   of several automatic test equipment (ATE) testers,
   a test and rework   several rework stations, and a local data base located
   work cell   in a computer separate from the testers. With
                             different board types entering the cell, the local data
                             base contains more control information, providing
                             routing direction for both boards and test programs to
                             the proper tester.
                                  Suppose there are nine different board types in
                             the product mix entering the work cell, which also has
                             three ATE and three rework stations. Each tester can
                             handle three board types. Any board may be repaired
                             at any rework station. As an individual board enters
                             the work cell, the data base must identify the board
                             type and route the board to the tester currently
                             equipped to handle that type. If a tester is not
                             available, the board must be held in a local buffer
                             storage area until the tester is ready for it. The system
                             must also verify that the correct test program has been
                             downloaded to the tester. At test completion, repair
                             information (if any) plus production data, such as time
                             and which tester was used, are stored in the data
                             base. This information may then be retrieved by any
                             of the three rework stations when the board arrives
                             at it.
                             Knowledge bases: The knowledge-based system
                             should perform four separate functions:

                             1. Scheduling testers.
                             2. Downloading test programs to individual testers.

3. Routing individual test and repair information, e.g., to guide light pointer at the rework station to the components to be replaced.
4. Providing summary reports of the test performances, process yields, etc.

Feasibility: Mixed chaining.

3. Intelligent PCB tester

Framework: Most existing ATEs run a structured series of tests, which may take many hours per PCB. In addition, the ATE does not always find all failures, because all the possible conditions are not tested. Finally, the ATEs are often only capable of isolating a string of components. As a result, in addition to the malfunctioning component, several functioning components would be replaced in vain.

Programming: OPS-5 and Assembler, or Prolog.

Feasibility: A knowledge-based system could build the test program based on test specifications and schematic diagrams for the PCBs. It should also run the actual test equipment in order to change the test methodology according to the test results. Testing and test result interpretation, especially in the case of analog circuits, is a complex task, and developing an autonomous expert system for it is not easy.

Inference: Backward chaining and truth maintenance.

4. Generation of repair instructions for printed wiring (PWB) assemblies

Framework: The goal is to combine knowledge bases about the PWB assembly process and design verification rules in order to generate on-line repair instructions.

Feasibility: Feasibility is conditional on the acquisition of the two above-mentioned knowledge bases.

Inference: Forward chaining.

5. SMT repair advisor

Framework: The SMT rework method varies with the type of defect, and component characteristics may not be known to the repair technician who needs advice.

Knowledge bases: SMD component characteristics (e.g., packaging, outline, electrical properties), defect types, and rework methods.

Programming: Expert system shell.

Feasibility: There have been developments of simple rule-based expert systems, with good user interfaces.

Inference: The inference selects the lowest-cost rework approach.

12.6. INTEGRATION

12.6.1. Configuration

1. Configuring system
   according to
   specifications

Framework: In many cases, the PCB configuration in the rack depends on the customer order or specifications. Quite often, the configuration is done simply by using manuals and/or forms or just modifying the configuration of a standard package, or that of an earlier delivery. Manually done, the task tends to be tedious and error prone. Often, the configuration cannot be completed in one go, because of incomplete specifications or expected changes, which increases the throughput time.

Sometimes, the system is to be installed outside the manufacturing site into a larger system, e.g., a control system of a lift or automatic warehouse. In this case, it is important, for easy installation and testing, to have a standard configuration methodology. This means, for example, that whenever signals are present in a configuration, they are always placed in standard connectors and pins. At least, the signals should appear in a standard order. Requirements and restrictions such as these make manual configuration inadequate in large-scale production.

Knowledge bases:
- Available system features.
- Resources (hardware, software).
- Configuration rules (how to configure).
- Valid configurations (what can be delivered).

Programming: Configuration rules are often relatively simple; thus Prolog or OPS-5 might be adequate. KEE or similar shells should be used for complex tasks.

Inference: Forward chaining.

12.6.2. Final Testing

1. Diagnostics of
   equipment consisting
   of several PCBs

Framework: Equipment or module diagnostics consists of carrying out tests and measurements, in view of detecting, locating, and identifying failures at the component, link, and/or least-repairable-unit level.

Knowledge bases: The required knowledge bases include:

- I/O descriptions of all components and links.
- Description of all interconnections, either physical or through functional and causal relations (using CAD).
- Failure mode and effects analysis (FMEA), in procedural form.
- Failure logs.
- Test selection and testability procedural knowledge.
- Failure detection, location, and diagnosis procedural knowledge.
- Maintenance and repair procedural knowledge.

Inference: Although diagnosis typically involves backtracking reasoning, diagnostic expert systems usually have mixed inferencing, in view of procedural and measurement constraints. Inference engines are now customized for heuristic search procedures, each specific to one equipment type; early feasibility tests, however, rely on standard expert systems shells.

2. Knowledge-based test program generation

Framework: Using design and process data, test vectors should be selected in a knowledge-based way, thus enhancing current algorithms which are highly constrained. The task is important, as it is estimated that test generation represents 24–34% of the total design time.

Knowledge bases: CAD layout, and electrical characteristics.

Programming: LISP.

Feasibility: The problem is, in general, quite difficult.

Inference: Mixed chaining with constraint propagation and triggers.

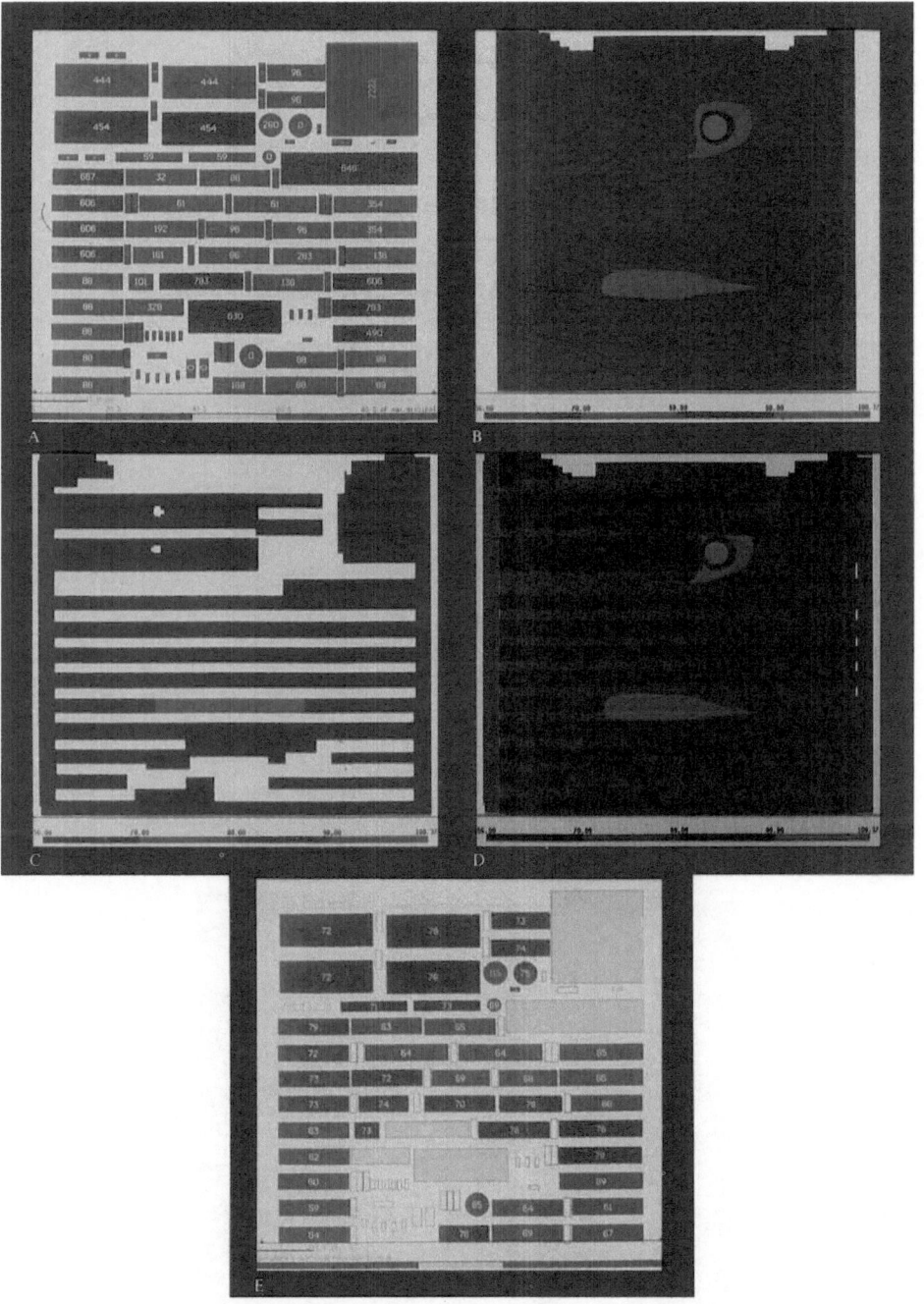

Figure 90. AOI-based processing, for thermal PCB imaging[84]: (A) gives component types, (B) the substrate layout, (E) the IR emissivity corrections, (D) the thermal image after thresholding, emissivity corrections, and pseudo-color coding, and (B) the same as (D) after subtracting the background (C).

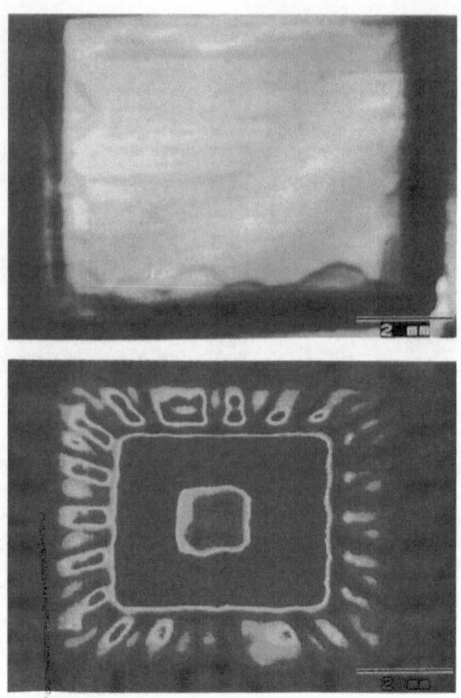

Figure 106. Acoustic images of a leadless chip carrier: Top image through the lid shows uneven bonding around the edges; bottom image through the base (pseudo-color coded) shows partial bonding and uneven lead adhesion.

Figure 131. Registration through rulers. (Courtesy AdeptVision Inc.)

Part II

Vision Algorithms for Electronics Manufacturing

In Part II, the various algorithms used so far are described in detail. The organization of the chapters in Part II is illustrated by the image understanding architecture in Figure 118.

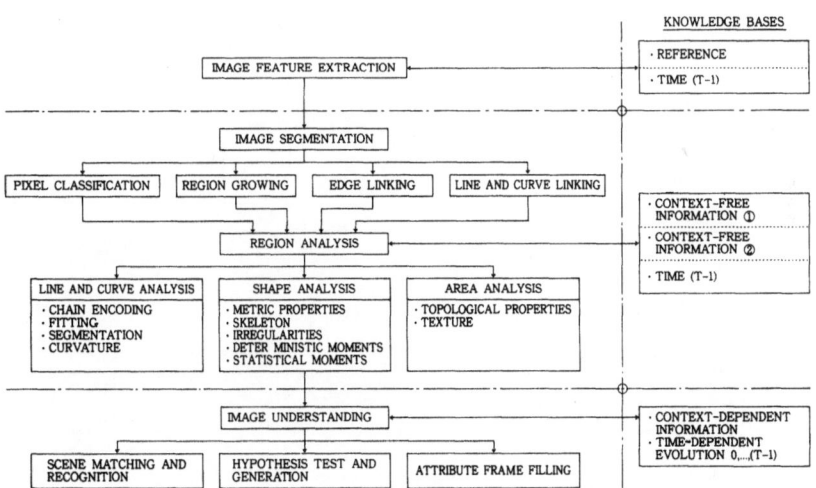

Figure 118. Typical, but not unique, organization of the basic image processing functions and examples of algorithms, with the required knowledge/fact bases.

Chapter 13

Image Quantization and Thresholding

13.1. ALGORITHM QUANT-1: QUANTIZATION

The quantization process transforms, for all pixels in the image, the corresponding analog sensor output, s, into a digital value, $d(s)$. Because of this quantization, each digital value will correspond to a range of values s. The number of quantization levels $d_0, d_1, \ldots, d_n$, where the d_i are the values $d(s)$ may assume, may itself be relatively small or large.

The amplitude density function $p(s)$ is the probability density of the output s over the range of quantization levels (see Figure 119).

The issue is then to select the quantization levels d_i in such a way as to minimize the error resulting from this process and/or to enhance those signal value ranges which are the most valuable for the inspection tasks.

Because the amplitude density function is unpredictable for most images, the typical quantization profiles are based on either (Figures 120 and 121) linear quantization, power law (gamma) quantization, or logarithmic quantization, or are made dependent on surface features.

In addition, one must select the number of quantization levels, and several criteria are possible:

1. The number n quantization levels should be proportional to the signal-to-noise ratio (SNR) of the image source; if a video or CCD camera is used, this gives approximately 1 bit per 6 db SNR, or $n = 7$ bits for cheap 42-db SNR cameras, and $n = 9$ bits for 54 db.

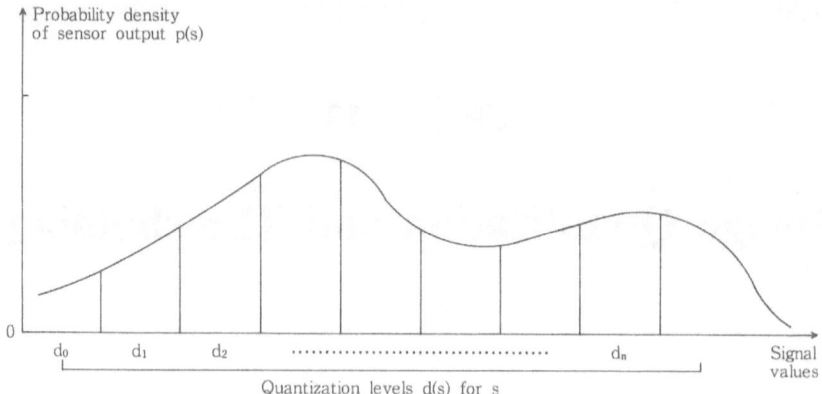

Figure 119. Image quantization levels.

2. Maximize the visual discrimination of luminance steps, by letting the number of quantization levels n be proportional to the logarithm of the bandwidth $(B_{max} - B_{min})$:

$$(\log B_{max} - \log B_{min})/n = 0.02$$

where the last value is the one the human eye achieves. So, for example, for a $50:1$ dynamic range $n = 197$ levels or 8 bits are used.

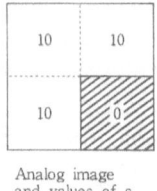

Analog image
and values of s

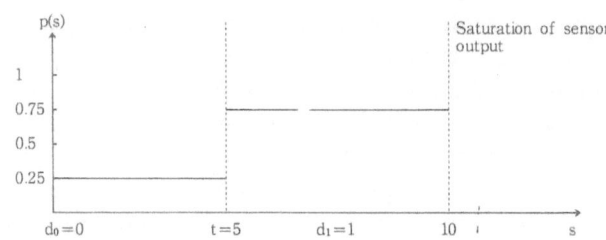

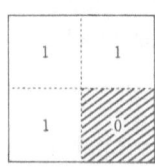

Digital image
and values of d(s)

Figure 120. Binary image and its quantization.

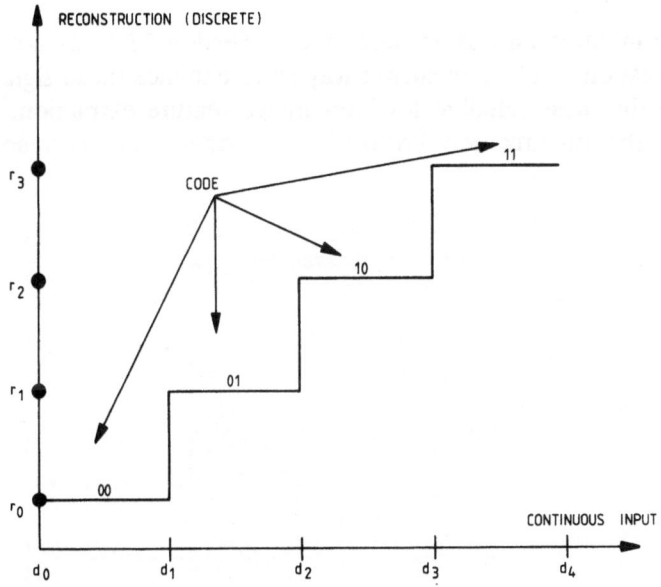

Figure 121. Quantization of an image and codes for each level: d_i, decision levels; $d_i < r_i < d_i + 1$; r_i, reconstruction levels.

3. Minimize contouring, which is visible under 7–8 bits, even on displays with limited dynamic range.

Regarding the selection of a quantization profile that is dependent on surface defects, the simple approach is to achieve through a surface treatment (coating, painting) and stable illumination, a good gray level and/or color separation between physically separate features. The quantization levels must then obey this gray level/color separation.[101]

13.2. ALGORITHM LUT-1: LOOKUP TABLE (LUT) TRANSFORMS

If the initial quantization described in Section 13.1 has not selected the quantization levels d_i in such a way as to enhance those signal values which are the most valuable for later image feature extraction, one can reallocate the quantization levels d_i by a one-to-one correspondence

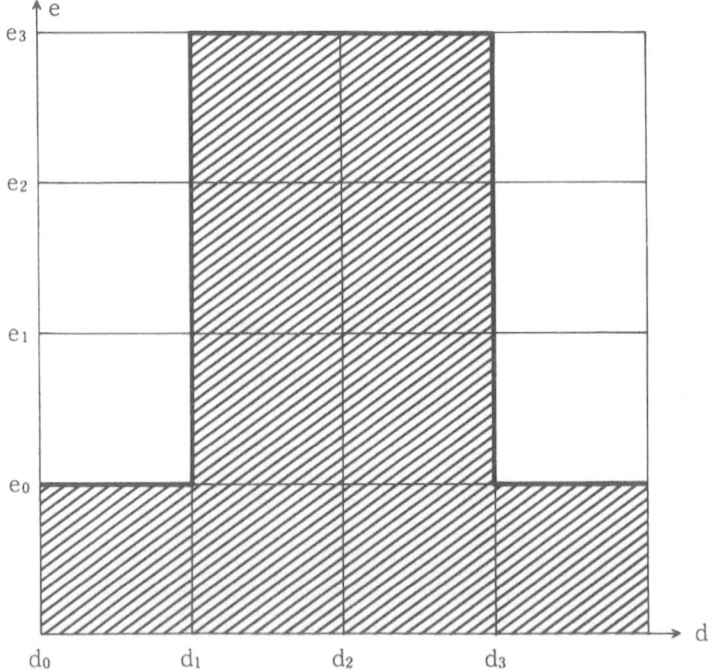

Figure 122. Example of a lookup table (LUT transform for $n = 3$ quantization levels); this one implements a band-pass filter ("slicing") transforming the analog image s into a binary image $e(s)$ with the highest e-quantization value if $d(s) = d_1$ or d_2.

LUT, without changing the boundaries or the number of the original quantization intervals (d_i, d_{i+1}), so that the digital image $e_i = \text{LUT}(d_i) \in \{d_0, \ldots, d_n\}$ becomes:

$$e(s) = \text{LUT}[d(s)]$$

Especially, the LUT can implement high-pass, low-pass, or band-pass filters (see Figure 122). Because of their ease of implementation, such LUT transforms are very fast and widely used.

13.3. ALGORITHM PSEUDO-1: PSEUDO-COLOR DISPLAY

To highlight defects or features in a gray-level image on a color display unit, one may want to assign a color value to the quantization levels d_i, in addition to the reallocated quantization level.

Typically, most color display units admit three color channels (e.g., R: red, G: green, B: blue, or equivalent); thus, three lookup tables, $\text{LUT}(k)$, $k = 1, 2, 3$, must be defined, and the pseudo-color transforms become:

$$e(i, k) = \text{LUT}(k)(d_i) \in \{d_0, \ldots, d_n\}^k$$

where the superscript k indicates that the quantization level $e(i, k)$ is assigned to the color display channel k.

13.4. ALGORITHM THRESH-1: FIXED THRESHOLD AND BINARIZATION

Thresholding is a special case of quantization, with $n = 2$ levels, d_0 and d_1. The result is a binary picture, where d_0 is one level, and d_1 the other; typically, $d_0 = 0$ and $d_1 = 1$.

As for binary images, quantization profiles as described above are meaningless; the issue is only to select the threshold t such that

$$d(s) = \begin{cases} d_0 & \text{for } s < t \\ d_1 & \text{for } s \geq t \end{cases}$$

The threshold above is applied throughout the area of interest (AOI) to which it is valid, and occasionally on the full image.

Although binary images are useful, it should be remembered that real life is not just black or white. Better performances are obtained by keeping application-specific gray-level image quantization schemes as long as possible. Binary thresholding usually implicitly requires increased spatial resolution, so that edge and other singular pixels comprise a smaller fraction of the total number of pixels, thus reducing the overall effect of edge variations. While increasing spatial resolution reduces the variation of measurements on binary images, gray-level processing performs better even at high spatial resolution.

13.5. ALGORITHM THRESH-2: THRESHOLD SELECTION BY THE DIFFERENCE HISTOGRAM

The selection of a threshold for binarization of an area of interest cannot be made from a fixed constant, because object properties as well as lighting and sensor fluctuations make it necessary to characterize each image separately.

Instead of binarizing the image in a straightforward way, one approach is to convert the image to any number of discrete levels by dividing its gray scale into segments, and setting a threshold t at the maximum value of the difference histogram in each segment.

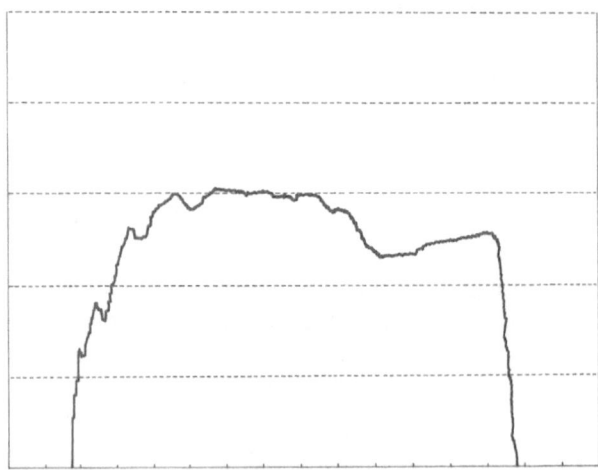

Figure 123. Example of a difference histogram; the gray levels are along the horizontal axis, and the frequencies of each gray level are along the vertical axis.

Let the range of gray levels be $0, \ldots, N$, in the image $x(i, j)$; $i = 1, \ldots, I; j = 1, \ldots, J$.

1. Calculate the gradient image $y(i, j)$, e.g., by a 4-adjacency estimate:

$$y(i, j) = 4x(i, j) - y(i, j - 1) - y(i, j + 1)$$
$$- y(i - 1, j) - y(i + 1, j)$$

2. Calculate the difference histogram $z(l)$ (Figure 123), which counts the number of positive or negative gray-level differences of amplitude l in $y(i, j)$, for $l = -N, \ldots, 0, \ldots, N$, and $i = 1, \ldots, I; j = 1, \ldots, J$.

3. Select the threshold t at any of the following alternative values:

$$t_1 = \tfrac{1}{2} - \{\text{Max}[z(l), 0] + \text{Min}[-z(l), 0]\}/(I \cdot J)$$

$$t_2 = \text{Max}[z(l), 0]/(I \cdot J)$$

Chapter 14

Geometrical Corrections

Correcting the geometrical distortion of an image is frequently necessary. Such distortion can, for instance, include the perspective distortion caused when the object is imaged from a lateral side, and the bending distortion caused, for example, by the scanning system.

When the object is moved through the depth of field, orthographic image acquisition lenses may keep the image size constant.

14.1. ALGORITHM GEOM-1: GEOMETRICAL CORRECTION

Geometrical distortions are defined by equations which transform a coordinate system (x, y) bearing no distortion to another coordinate system (u, v) with distortion.

Transformation equations include:

1. Linear/affine transforms

$$\begin{bmatrix} u \\ y \end{bmatrix} = \begin{bmatrix} a & b \\ c & d \end{bmatrix} \begin{bmatrix} x \\ y \end{bmatrix} + \begin{bmatrix} u_0 \\ v_0 \end{bmatrix}$$

2. Quadratic conformal transforms

$$u = ax + by + c(x^2 - y^2) + 2dxy + u_0$$
$$v = -bx + ay + 2cxy - d(x^2 - y^2) + v_0$$

3. Hermat transform (rotation, zooming, shift)

$$\begin{bmatrix} u \\ v \end{bmatrix} = \begin{bmatrix} a & b \\ -b & a \end{bmatrix}\begin{bmatrix} x \\ y \end{bmatrix} + \begin{bmatrix} u_0 \\ v_0 \end{bmatrix}$$

4. Projective transforms

$$u = (a_1 x + a_2 y + a_3)/(a_7 x + a_8 y + 1)$$

$$v = (a_4 x + a_5 y + a_6)/(a_7 x + a_8 y + 1)$$

This last transform moves a viewing axis without changing the viewing point.

14.2. ALGORITHM GEOM-2: INTERPOLATION

Let f and g represent an image with no distortion and the same image distorted by the transformations h_1 and h_2, respectively. Thus, $g(u, v) = f(x, y)$, because the feature which should appear at point (x, y) will actually appear at point (u, v). When computing the point

$$(\alpha, \beta) = [h_1(x_0, y_0), h_2(x_0, y_0)]$$

of a distorted image corresponding to each point (x_0, y_0) of the original image f, (α, β) does not always coincide with the lattice point of g because the values α and β are not usually integers. Accordingly, some

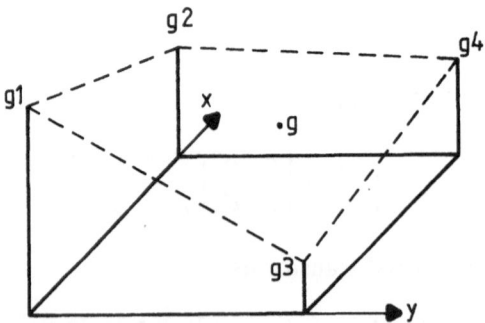

Figure 124. Four-point bilinear interpolation, as also used in software-based "zooming." $g(x, y) = (g2 - g1)x + (g3 - g1)y + (g4 + g1 - g3 - g2)xy + g1$.

form of interpolation is necessary, and typical procedures include:

1. 4-point bilinear affine interpolation (as used in software zooming) (Figure 124).
2. 9-point quadratic affine interpolation.
3. Nearest neighbor or maximum/minimum value of four neighbors.

14.3. ALGORITHM GEOM-3: SUBPIXEL EDGE INTERPOLATION

The profile across an edge will never be perfectly binary but will show a gray-level variation. The edge location procedures of Chapter 16, however, only give the edge location with a resolution equal to the pixel dimensions. The question is, "Where is the edge?" There are numerous ways to compute the edge location, several of which are:

- Compute the pixel location halfway between the maximum and minimum gray values across the edge.
- Compute the pixel location of the highest rate of change using a first difference calculation, or use the average location of all points having the same first difference as the highest rate of change.
- Fit a least-squares polynomial through the points, and then find the location of the highest-derivative point.

All of these methods will possibly yield an edge location that is not an integer and are therefore called subpixel methods.

14.4. ALGORITHM CALIB-1: SENSOR CALIBRATION

Sensor calibration is the determination of the correspondence between points in the camera image and points in physical space. Only when the sensor is properly calibrated can its image master pixel coordinates be translated into real object locations.

The assumptions made in sensor calibration are:

1. The lens behaves as an ideal thin lens.
2. The image pixel matrix is aligned with the external coordinate system.
3. The object surface is perfectly diffusely reflecting.

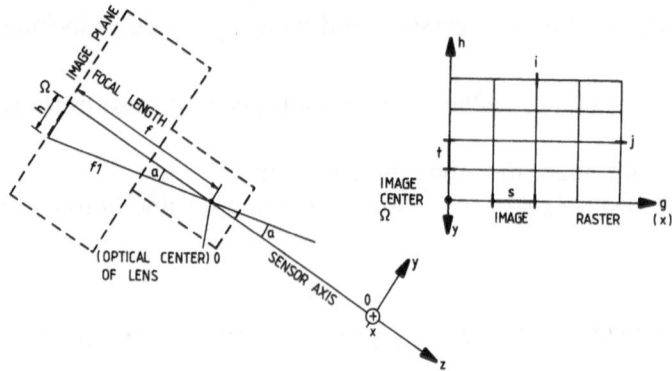

Figure 125. Projection of the sensor lens calibration onto the y–z plane; the h–g image raster coordinates are also shown.

Corrections for distortion, modal separation, master rotation, and sensor/light source misalignment are given in Ref. 102.

Sensor calibration involves the calculation of the following parameters (see Figs. 125–127):

- Pixel spacing and camera coordinates:

$$g = i \cdot s \qquad x = (z/f) \cdot s \cdot i$$
$$h = j \cdot t \qquad y = (z/f) \cdot t \cdot j$$

- Distance d_1 of the illuminated object P to crossing point K:

$$d_1 = A \cdot j/(1 - B \cdot j)$$

with

$$A = d \cdot (t/f) \cdot (1/\sin \beta)^2$$
$$B = (t/f) \cdot (\cos \beta/\sin \beta)$$

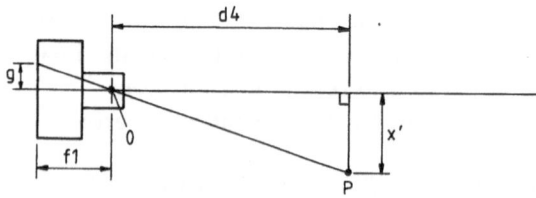

Figure 126. Projection from above the sensor/lens calibration onto the x'–y' plane with relations between external sensor coordinates (x, y, z) and image raster coordinates $(h$–$g)$; P is the illuminated point on the object in the $x' = 0$ plane.

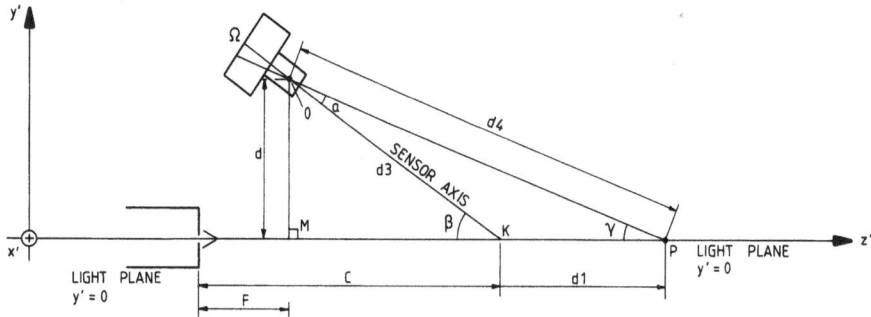

Figure 127. Side view projection onto the $y'-z'$ plane of the sensor/lens calibration.

A corresponds to the lens magnification.
- Distance z' of the object P to the light source:

$$z' = (A \cdot j)/(1 - B \cdot j) + C$$

Note that the constant B corresponds to the vanishing point where z' is infinite.
- Coordinate x':

$$x' = D \cdot i \cdot [E + (z' - F)^2]^{1/2}/[1 + (h/f)^2]^{1/2}$$

where $D = s/f$ is the magnification in the x' direction, $E = d^2$ is the separation between the camera and the light plane, and f is the z' coordinate of the optical center of the lens (it is zero if the origin of z' is at the optical center of the lens).

To measure the coefficients $A-F$, one can invert the above formulas, by transforming observed raster coordinates (i, j) of known points to the real coordinates (x', z'). These calculated values (x', z') can then be compared with the physical coordinates actually observed. A goodness-of-fit measure, ε_0, can then computed in the observed image coordinates between the observed coordinates (i, j) and the nominal ones (i^0, j^0):

$$i = (x'/D)/\{[E + (z' - F)^2]^{1/2}/[1 + (j \cdot t/f)^2]^{1/2}\}$$
$$j = (z' - C)/[A + B \cdot (z' - C)]$$
$$\varepsilon = \Sigma [(i - i^0)^2 + (j - j^0)^2]$$

The parameters are then selected to minimize ε.

Chapter 15

Image Registration and Subtraction

Image registration is the process which carries out the alignment of an input image with a template thereof, within a search region. Subsequently, registration also gives the location and displacement where the optimal matching is obtained. Deviations from matching are revealed by, for example, image subtraction.

15.1. REGISTRATION PROBLEMS

15.1.1. Registration Subproblems

There are several registration problems for which external information is required to establish that points, features, or other items in two different image frames M-i and M-j are actually identical in the physical sense. These registration subproblems P-k are defined for frame models M-i.

In the remainder of this section, we assume the sensor to be fixed. Otherwise, it is necessary to incorporate the transformation relating two different views of a 2-D or 3-D scene as a function of the displacement of the camera relative to the scene between the taking of the two views.

The registration compensates for shift in x, y, z directions, including rotation and eccentricity; scale change; optical or other geometrical distortions (e.g., mechanical vibration); incomplete observations (partially masked areas); and known movements/kinematics. All of the above are related to geometry, and thus quite critical for any measurement/sorting task.

All registration subproblems, P-1, P-2, P-3, are numerically very intensive. An important point is that all the remarks made here are based on the assumption that there is only one (segmented) object per frame. The complexity of these subproblems, in the *ideal* case (free from measurement errors), can be *approximated* as varying as follows:

P-1: Proportionate to number of shape features (Chapter 19); varies as fourth power of number of offset pixels *and* as at least third power of number of object types.

P-2: Proportionate to number of shape features *and* varies as second power of chain code length (Algorithm Chain-1 or Moments-2).

P-3: Proportionate to number of edge points *and* to cluster sizes used.

This complexity is further increased by the errors at the M-3 and M-4 levels, due to the low-level image processing algorithms used (thresholding, edge point extraction, linking, etc.).

The above is basically valid whether the registration reference frame M-0 is available as a vectorized CAD file input, a rasterized CAD file

Table 38. Registration Subproblems P-1, P-2, and P-3 and Relation to Image Frames M-*i*

Subproblem	M-*i* (input)	M-*j* (reference)	Type of procedure	Output
P-3	M-3	Basic edge element shapes, lengths (edge primitives) (M-4)	Geometrical proximity between two clusters of points	Registered M-3
P-2	M-2	Singular shapes (shape primitives) (M-5)	Shape proximity, e.g., chain code correlation or syntactic filter	Registered M-2
P-1	M-1	M-0	(a) Recognition or verification of object type (b) Estimated parameters of geometrical transformation between identical shape primitives	Registered M-1

input, or a rasterized "teaching-by-showing" physical image with graphic overlay ("mouse," digitizing table) (Algorithm Templ-1). The differences between these three cases are only in the implementation of the P-1 software and in data access procedures (Algorithm CAD-1).

15.1.2. Registration Procedures

The basic registration procedures for each of the subproblems P-1, P-2, and P-3 are given in Table 38 in connection with the subproblem definitions. Other classes of algorithms or procedures used in, or relevant to, the proposed registration problem decomposition are listed in Table 39.

15.1.3. Registration Errors

Registration errors result from four causes:

1. Sensor measurement errors, including related preprocessing (see Section 1.5.1).

Table 39. Other Registration Procedures

Subproblem	Classes of procedures
P-1	Syntactic recognition[a] (Algorithm Fuzzy-1) Probabilistic relaxation[b] (Algorithm Label-2) Geometrical transformation estimates between point sets (Algorithm Geom-n) Local optical flow[c]
P-2	Chain code correlation (Algorithms Chain-1, Moments-n) Syntactic filters[a] (Algorithm Fuzzy-1) Graph matching[d,e] (Algorithm Bridge-1)
P-3	Dynamic cluster analysis[f] Graph matching[d,g] (Algorithm Bridge-1) Fast image comparison by simplified rank correlation[h] (Algorithm Reg-2) Pattern matching and parameter estimation Dynamic programming
P-4	Geometrical transformation estimates between point sets, initialized with static point positions

[a] Ref. 103. [e] Ref. 106.
[b] Ref. 104. [f] Ref. 107.
[c] Refs. 75, 94, and 105. [g] Ref. 108.
[d] Ref. 77. [h] Ref. 109.

2. Registration algorithm errors (P-1 through P-4).
3. Errors in registration reference frame (denoted M-0) (see Section 1.6.3).
4. Curved surfaces: Curvature makes the reflection vary even on a perfect part; registration errors remain even if the vision system analyzes the direction of variation of light at each point.

15.1.4. Registration versus Scan Type

The issues here are only:

S-1: Size of the window W used to carry out the registration process P-k:
Example 1: $W = 1 \times 1024$
Example 2: $W = 512 \times 512$

S-2: Sequencing of the scan, digitizing, registration, and lower-level preprocessing required to reach M-3 in the window W (see Section 1.3).

The scan is started when an object is first detected and when it can be assumed that most of it is seen.

According to the choices made for S-1 *and* S-2, different systems or architectures are required to obtain adequate results. As shown above, large windows W will affect the processing time and complexity of P-3 and P-2. Insufficient pipelining or unused scanning, plus preprocessing time, will result in the requirement for much faster central host hardware. Pipelining allows for concurrency.

The performance criteria to use for comparison of alternate scan types in registration are:

- Instruction rate per pixel.
- Memory occupation (transient and average).
- Parallelism.
- Off-line precomputations.

15.1.5. Avoiding Registration

All the listed sensor-related measurement errors that occur in vision applications will often remain to some extent, although some may be partially compensated for by delicate customer- and application-specific engineering[110] covering strobing, flash light (see Section 1.4), collimated light (homogeneous), mechanical position gauging, movement servoing (see Section 6.1), adaptive (locally adapted) image processing (see, e.g., Algorithm Thresh-2), and inspection sampling rate (see Section 1.7.1).

It is therefore easy, and tempting, to believe that registration can be avoided, or partially compensated for, in vision-related *metrology* and in measurement-related *sorting* tasks. Analysis of the registration steps (P-1 through P-4) demonstrates, however, that this is not so.

It is even less so if the vision-based metrology and/or sorting system is conceived as an off-the-shelf component, for which customization should be limited. Because there will be sensor measurement errors, there will always be residual registration errors to be compensated for in vision-related metrology or sorting tasks.

Typical compensations for registration are:

- Human/robot positioning of the item with an accuracy significantly higher than the dimensional quality control acceptance limit.
- Using a second vision system to carry out, with a mechanical handler, the correct alignment and positioning of parts.
- High-accuracy noncontact position gauging (e.g., by an inter-ferometric or holographic grating on the objects or by a cus-tomized battery of fiber optic position sensors) (see Chapter 3).

15.2. ALGORITHM REG-1: CORRELATION

Correlation (Ref. 111) can be used, but the calculation time is very long, even after FFT. Let $x(i, j)$ be the original image, $t(i, j)$ the template, and $y(i, j)$ the subset of x of same size as t, centered at location (l, m) within x (Figure 128).

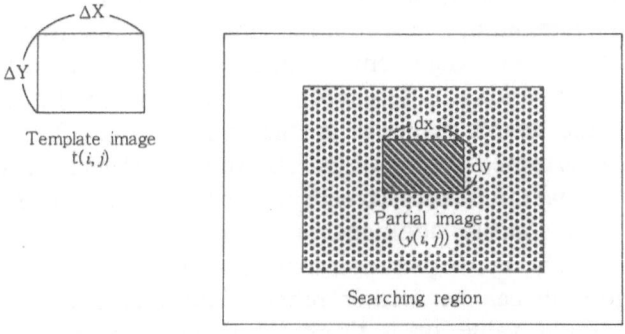

Figure 128. Correlation between template and input image.

The normalized cross-correlation coefficients c_{lk} between the template image $t(i, j)$ and the partial image $y(i, j)$ are calculated using

$$c_{lm} = \sum_{i=1}^{I} \sum_{j=1}^{J} \frac{[t(i, j) - T][y(i, j) - Y_{lm}]}{(s_t \cdot s_y)^{1/2}}$$

where I, J is the size of the template image, and

$$T = \frac{1}{I \cdot J} \sum_{i=1}^{I} \sum_{j=1}^{J} t(i, j) \qquad \text{for } t \text{ centered at } (l, m)$$

$$Y_{lm} = \frac{1}{I \cdot J} \sum_{i=1}^{I} \sum_{j=1}^{J} y(i, j) \qquad \text{for } y \text{ centered at } (l, m)$$

$$s_t = \sum_{i=1}^{I} \sum_{j=1}^{J} [t(i, j) - T]^2$$

$$s_y = \sum_{i=1}^{I} \sum_{j=1}^{J} [y(i, j) - Y_{lm}]^2$$

The template window is moved across to another location (l, m), until a maximum of c_{lm} is achieved.

15.3. ALGORITHM REG-2: SEQUENTIAL SIMILARITY DETECTION

The basis of sequential similarity detection is to use the sum of the differences in absolute value[112,133] between each pixel in the input and template images. When registration is perfect, the sum of these residual differences is a minimum, and zero if the difference is free from noise and geometrical distortion. Alignment is attained at the location where the minimum sum of residual differences is obtained, when the template is moved over the search window in the input image. The search is immediately interrupted and restarted elsewhere in the search window if the sum of residual differences exceeds a specified threshold. Because the procedure requires mostly additions only, and because the search can be curtailed selectively, the registration time is small.

The algorithm can be further refined into variants, with either a constant threshold value or a threshold linearly increasing with the number of search locations. Prediction can also be used to truncate the useless searches earlier.

15.4. ALGORITHM TEMPL-1: PHYSICAL TEMPLATES

Registration can sometimes be implemented the most efficiently by having physical templates attached to the object being imaged or physically scanned over it. This includes holographic templates, X-ray templates in PCB drilling, and drill pilot holes.

15.5. ALGORITHM CIRCLE-1: CIRCLE FITTING

If one imagines a circular object in the area of interest and locates its edge points along integer-valued pixel boundaries, one can take those edge points and use them in a least-squares circle fitting calculation, where x_c, and y_c are the center coordinates, and r the radius:

$$x_c = -A/2$$
$$y_c = -B/2$$
$$r = (x_c^2 + y_c^2 - C)^{1/2}$$

and where A, B, and C are the solution of the following least-squares equation:

$$\begin{bmatrix} S(x^2) & S(xy) & S(x) \\ S(xy) & S(y^2) & S(y) \\ S(x) & S(y) & N \end{bmatrix} \begin{bmatrix} A \\ B \\ C \end{bmatrix} = \begin{bmatrix} -S(x^3) - S(y^2x) \\ -S(yx^2) - S(y^3) \\ -S(x^2) - S(y^2) \end{bmatrix}$$

where x and y are edge point locations, S is summation of the edge point pixel values, and N is the number of points on the edge of the circle.

Note that the x–y center and the radius of the circle are computed from measurements made along every edge point. The result computed will be a cumulative calculation based on numerous observations and measurements. In general, the accuracy of the measurements will increase as the square root of the number of points taken. Such subpixel techniques have been used to measure silicon wafers up to 6″ in diameter with 1-mil/pixel resolution cells, yielding registration accuracies of 0.2 mil.

15.6. ALGORITHM REG-3: MULTIPLE RESOLUTION ALIGNMENT

One efficient procedure is to use n different optical paths of different magnification, imaging the same area, and to store the resulting pixels as a quadtree with the lower spatial resolution at the top of the tree. The highest-magnification level provides a field of view that is large enough for gross alignment; the higher-resolution layers are then applied to more and more time-consuming registration algorithms, e.g., Algorithm Templ-1 first, then Reg-2, then Reg-1.

15.7. ALGORITHM OFFSET-1: OFFSET CALCULATION FROM LINE INTERSECTIONS FOR FAST ALIGNMENT

Algorithm Offset-1 is the basis of many automatic alignment subsystems such as those used in probers, dicing saws, pick and place systems, die bonders, wire bonders, screen printers, wafer handlers, laser trimmers, SMD placement, and low-resolution linewidth measurement.

This very fast alignment procedure consists of the following steps (see Figure 129):

1. Extract edges of the object to be aligned (by an Edge-n algorithm).
2. Calculate from the result of step 1, and object model knowledge, a coordinate axis or reference system attached to the object.
3. Project a pattern of lines or curves of known absolute coordinates onto the edge picture from step 1.
4. Calculate the absolute coordinates of the intersection points between the results of steps 1 and 3.
5. Calculate the offset between the results of steps 1 and 4 for generation of positioning commands.

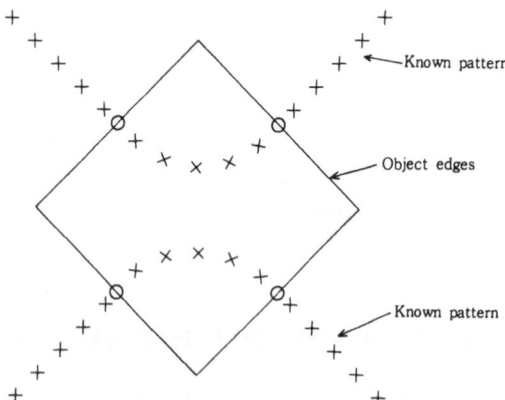

Figure 129. Offset calculation from the absolute coordinates of the intersection points (O).

15.8. ALGORITHM RULER-1: VALIDATION OF A GEOMETRICAL MODEL

When the inspection tasks consist in checking whether a set of spatial relationships are satisfied or not, geometrical model validation applies.[114] One example is checking the spatial relationships corresponding to pads and conductor traces in a PCB mask.

In the simplest case, the algorithm involves searching for edges or object boundaries, along finite directed line segments in the image; these segments are called "rulers."

This algorithm consists in defining a model, made of a number of line segments (straight or circular), whose boundaries correspond to gray-level image boundaries; a small number of attributes describe each line segment (Figure 130). The inspection is then a matching decision between the model and the line segment with attributes derived from the image to be checked. An example is given in Figure 131. The algorithm has the following steps:

Step 0: Training: The model of the object is made of:
 • A reference line l_0, with known absolute coordinate of origin, end, shape, orientation, and length.

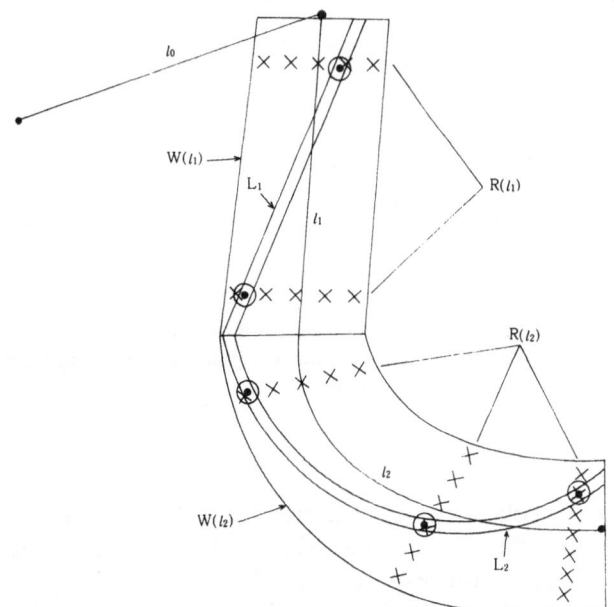

Figure 130. Geometrical model validation.

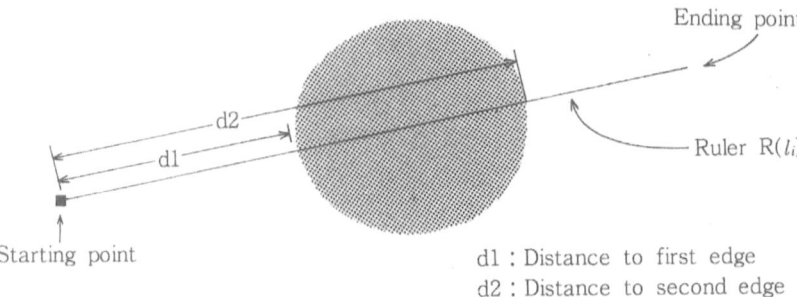

Starting point

Ending point

Ruler R(l_i)

d1
d2

d1 : Distance to first edge
d2 : Distance to second edge

Figure 131. Registration through rulers. (A color version of the top panel of this figure can be found at the end of Part I, after Chapter 12, facing page 205.) (Courtesy AdeptVision Inc.)

- A collection of feature lines l_i, with known relative coordinates for the origin, end, shape, orientation, and length with respect to l_0.
- A collection of area-of-interest or processing windows $W(l_i)$, each attached to and surrounding l_i symmetrically.
- A raster of straight search lines $R(l_i)$ in $W(l_i)$, typically orthogonal to l_i, and such that these search lines $R(l_i)$

intersect l_i in $W(l_i)$ at enough locations to find l_i in the window; the search lines $R(l_i)$ are called "rulers."
- A set of attributes for each line l_i:
  1. Length: fractional length l_i/l_0.
  2. Mean gray level: average gray level along l_i, with the corresponding standard deviation.
  3. Edge size: gray-level step between l_i and l_{i+1}.

Step 1. Line detection in image: Edges and lines in the image to be inspected are determined.

Step 2. Line calibration: The search lines $R(l_i)$ intersect the lines found in Step 1 at a number of points in $W(l_i)$. When linked together or when a best fit is found between them with knowledge of the reference shape from l_i, these points yield the observed lines L_i. The attributes of L_i are measured.

Step 3. Matching: It is assumed that the lines L_i are all found. Matching consists in grouping subsets of (L_i) to match these groups with subsequences of the same type from (l_i). The best match minimizes a global cost function, for example, a sum of the weighted difference terms between (L_i) and (l_i); the minimization is over all combinations; the weights are related to line attributes.

15.9. ALGORITHM SUBTR-1: IMAGE SUBTRACTION

Once two images are registered showing the same scene, two cases may occur: they are either identical or different, in terms of a dissimilarity measure, D.

To detect such a difference, if any, one may define the difference image $(x_1 \div x_2)$ between two registered images x_1 and x_2 of the same size as:

$$(x_1 \div x_2)(i, j) = x_1(i, j) - x_2(i, j)$$

where the right-hand side is the algebraic gray-level difference. It should be noted that the quantization range for $(x_1 \div x_2)$ covers $(-D, +D)$ if $D = \text{Max}|d_i - d_j|$ is the largest quantization level difference.

If, however, x_1 and x_2 are both binary images with quantization levels d_0 and d_1, defined by the same thresholds, then a related definition is the exclusive OR of x_1 and x_2:

$$\text{XOR}(x_1, x_2)(i, j) = \{d[x_1(i, j)]\} \cdot \text{XOR} \cdot \{d[x_2(i, j)]\}$$

where the right-hand side is the logical XOR value of the logical quantization levels.

One possible dissimilarity measure D is the number of pixels where:

$$(x_1 \div x_2)(i, j) \neq 0 \quad \text{or} \quad \text{XOR}(x_1, x_2)(i, j) = d_1.$$

Chapter 16

Edge and Line Detection

16.1. CATEGORIZATION OF ALGORITHMS

The discontinuities in the gray levels and/or color levels in an image are called edges and are basic parts of low-level image information. The detection of edges, and connected sequences thereof called lines, is a fundamental technique,[115] and the required algorithms are typically divided into the following categories:

1. Detection of edges and lines by applying local operators:
   - Differential type operators based on spatial differentiation; such operators have been developed by Roberts,[116] Prewitt,[117] Sobel,[118] Kirsch,[119] and Robinson.[120]
   - Model-fitting methods which assume an edge model in a small region and seek the estimated model parameters; such methods have been developed by Hückel.[121-124]
   - Statistical approaches which include high-pass filters and noise modeling.[125]
2. Enhancement of lines and curves, which eliminates noise from the results of the algorithms of type 1 and filters the elements, helping create long lines and curves:
   - Enhancement in image space, by checking the strength of neighbor elements perpendicular to the edge direction and eliminating weak elements; relaxation methods can be applied.[126]

- Enhancement in parameters space, based on clustering in the parameter space of the lines and curves to be extracted; this includes the Hough transform which expresses straight lines in polar coordinates ϕ, ρ, and obtains lines by counting in the (ϕ, ρ) plane, eventually accounting for edge direction.[127,128]
3. Connection of elements to extract smoother and longer lines:
   - Connections using graph search, where knowledge or constraints related to the sequence of elements to be detected are used in an evaluation function.[129]
   - Other methods.[115]

Many of the above methods involve applying either a neighborhood operation (types 1 and 2), which can be implemented efficiently by circular buffering (see Algorithm Buff-1 in Chapter 21), or relaxation labeling (type 3) (see Algorithm Edge-2 below).

16.2. ALGORITHM EDGE-1: FILTERING-BASED EDGE DETECTION

As mentioned in Section 16.1, one simple approach to edge detection is to apply a fixed weighting filter as a local operator to a fixed neighborhood around the candidate edge point and to apply a decision rule on the scalar product outcome (see Figures 132 and 133).

1. Select a fixed filter $F(m, n)$ with $m = -N, \ldots, 0, \ldots, N$ and $n = -M, \ldots, 0, \ldots, M$ over a fixed $(2N + 1) \times (2M + 1)$ window; the coefficients must be such that:

$$\sum_{m,n} F(m, n) = 0$$

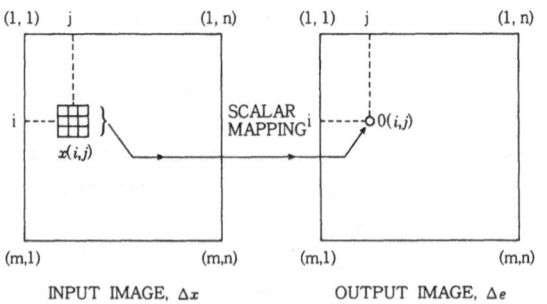

Figure 132. Filtering based on 3 × 3 local operations, with input image x and output image e.

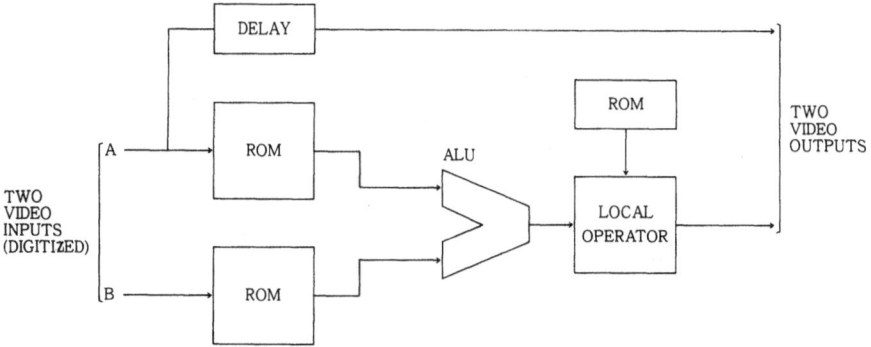

Figure 133. Preprocessor module capable of performing edge and line detection, image subtraction, expansion–contraction, thinning, and shrinking operations; the functions can be modified by changing the ROMs and altering the arithmetic and logic unit (ALU) controls.

2. At any candidate edge point location (i, j) in the image $x(i, j)$, calculate the scalar product/mapping:

$$e(i, j) = \sum_{m,n} F(m, n) \cdot x(i + m, j + n)$$

3. Pixel (i, j) is then an edge point if $|e(i, j)|$ exceeds some fixed threshold, which can be updated, e.g., w.r.t. edge contrast; $e(i, j)$ is then the edge intensity.

Typical filters F are described in Refs. 116–120; it should be noted that F can be made directional, so that step 3 resonates only if the edge has locally a specified direction or range of directions.

The filters can also be nonlinear, e.g., a Gaussian kernel:

$$F(m, n) = \exp[-(a_1 m^2 + a_2 n^2)/2\sigma^2]$$

in which case smoothing also takes place, so that the further operations from step 2 must be carried out on the gradient of the image smoothed by F. Multiresolution filtering corresponds to a range of values of σ in the definition of the Gaussian kernel F.

16.3. ALGORITHM EDGE-2: EDGE LABELING FOR LINE DETECTION

The goal is to track lines or boundaries, by assigning to each pixel on a line or boundary a consistent label giving the local slope or a no-edge

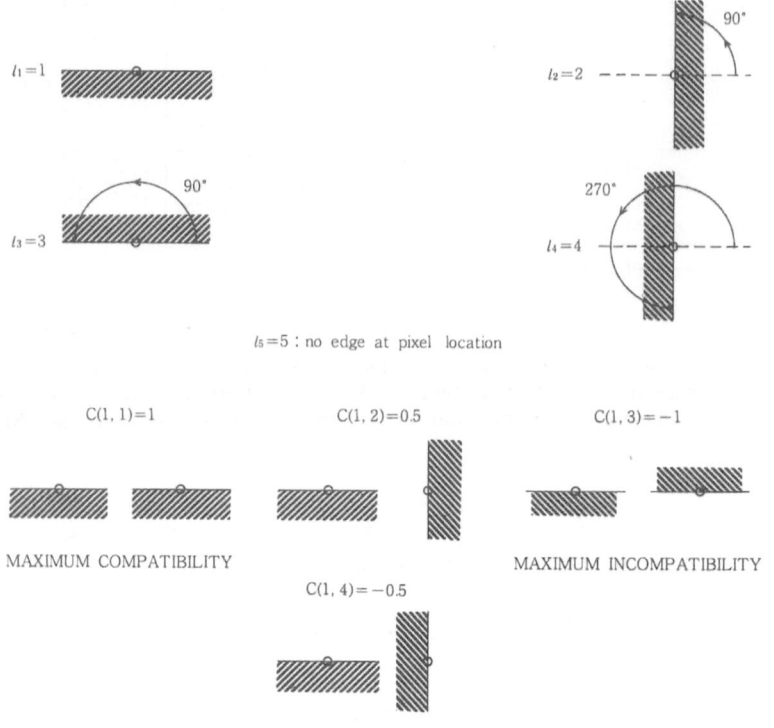

Figure 134. Edge labeling: specification of the edge labels l_i and compatibility coefficients $c(i, j)$.

label (Figure 134). This is clearly related to relaxation labeling (Algorithm Label-2).

1. Define the $N = 5$ possible pixel labels l_i, $i = 1, 5$, corresponding to four edge slopes of 0, 90, 180, and 270° and no edge ($i = 5$) at the pixel location (Figure 134); the label set can be increased or customized for specific cases.
2. Define the world model, to reflect the fact that boundaries consist of edge elements which line up approximately with each other's neighbors. To this end, one defines a set of compatibility coefficients of label l_i with neighbor label l_j: the values $c(i, j)$ are given in Figure 134, and $-1 < c(i, j) < 1$.
3. Carry out the initial tentative label assignment, by assigning, for example, a set of probabilities $p^0(l_i)$, $i = 1, N$, at each pixel,

based, for example, on the output of the edge detector, and such that:

$$\sum_{i=1}^{N} p^0(l_i) = 1$$

4. Poll the M neighbors to each pixel, and compute updating coefficients $q^k(l_i)$ for each label probability $p^k(l_i)$, based upon the probabilities of neighbor's labels $p^k(l_j)$, the compatibility coefficients $c(i, j)$, and the distance d_n to the neighbors:

$$q^k(l_i) = 1 + \sum_{n=1}^{M} \frac{d_n}{M} \cdot \sum_{j=1}^{N} c(i, j)p^k(l_j)$$

with $0 < q^k < 2$ expressing the compatibility between label l_i at the central pixel and the labels at the neighbor locations. Thereafter, the label probabilities are updated by:

$$p^{(k+1)}(l_i) = \frac{p^k(l_i)q^k(l_i)}{\sum\limits_{j=1}^{N} p^k(l_j)q^k(l_j)}$$

5. Iterate step 4, until convergence; usually only a few cycles are necessary.

16.4. ALGORITHM EDGE-3: SUBPIXEL EDGE EXTRACTION

Subpixel resolution (see also Algorithm Geom-3) allows the location of, for example, an edge with a positional accuracy smaller than the individual pixel size.

Another justification for subpixel accuracy is that it allows operation on a much larger field of view, with an accuracy comparable to that of a vision system having a "normal" resolution, but on a much smaller field of view than the subpixel system. The advantages of this capability are simplicity, higher speed, less camera stepping, and elimination of lighting difficulties associated with high magnification optics.

The basic steps involved are:

1. Resample each $n \times n$ window in the original image (e.g., an AOI) into a $(np \times np)$ subpixel higher-resolution image, with

$p > 2$; this is done by using an array processor to calculate the 2-D fast Fourier transform $(\text{FFT})^{130}$ of the $n \times n$ AOI, with a spatial quantization step of $z = (1/p)$ instead of $z = 1$; see also the Whittaker algorithm in Refs. 130 and 131.

2. Apply a linear edge detection algorithm (Algorithm Edge-1) in the spatial frequency domain of step 1, using the linearity of the resulting Fourier transform with respect to the filter function.

3. Apply the interpolation Algorithm Geom-3 at step 2 in the spatial frequency domain.

4. Apply the inverse FFT to step 1 to get the subpixel edge locations.

16.5. ALGORITHM EDGE-4: PYRAMIDAL EDGE DETECTION

When the image arrays $x(i, j)$ are very large and square,

$$i = 1, \ldots, N \quad j = 1, \ldots, N \quad N = 2 \cdot \exp p$$

one approach is to assume several visual resolution levels organized into

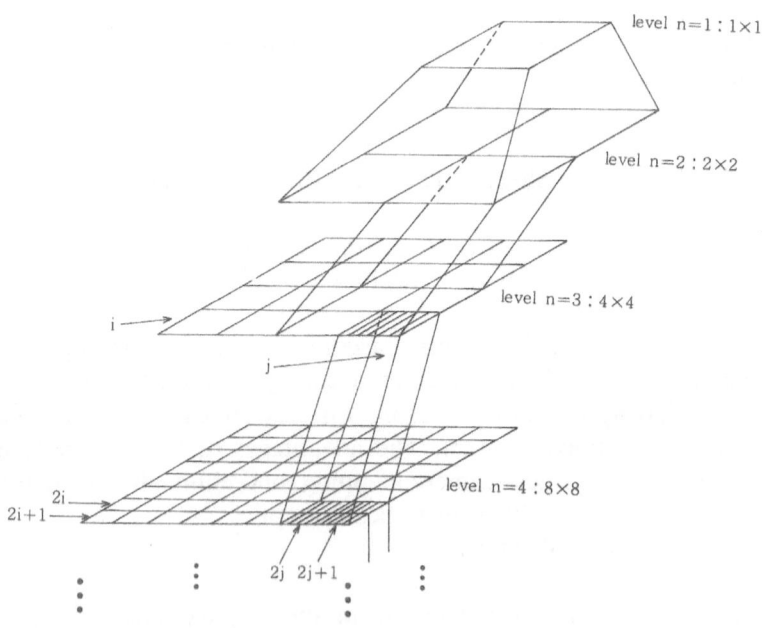

Figure 135. Image pyramid structure over $n = 4$ levels, with corresponding image subarrays.

a pyramid structure of two-dimensional arrays $y(i, j, n)$, $n = 1, \ldots, p$ (see Figure 135). This mimics a selective attention or focusing perception. The dimensions of the arrays double at each level n, until $y(i, j, n)$ is identical to $x(i, j)$ for $n = p$.

The high-level arrays $y(i, j, n)$, with n small, should therefore help detect the largest edges, if edge detection is applied to them. The low-level arrays (n large) give detailed edge positions; by using high-level results, implementation is possible for large values of N, in combination with varying spatial resolutions and selective attention.

In the algorithm Edge-4, a level n will be examined for edges; all edges found in the input array INP at level n will be indicated in the output array OUTP at level n. Edge-4 calculates edges at each pixel according to some detection operator BNDRY (see Algorithms Edge-1, -2, and -3). GET and STORE access pixel values. If the BNDRY filter output or edge strength is larger than THRESHOLD, the edge will be REFINED recursively.

```
PROCEDURE Edge-4-1 (inp, outp: pyramid; n: integer);
VAR i, j: integer
BEGIN
   FOR i := 0 to 2**n - 1 DO BEGIN
   FOR j := 0 to 2**n - 1 DO BEGIN
   store (outp,n,i,j,bndry(inp,n,i,j));
   IF get(outp,n,i,j) > threshold THEN
      Edge-4-2(inp,outp,n,i,j);
   END;
   END;
END;
```

```
PROCEDURE Edge-4-2 (inp, outp: pyramid; k,i,j: integer);
VAR di,dj: integer; temp: real;
BEGIN
   IF k < L THEN BEGIN
   FOR di := 0 to 1 DO BEGIN
   FOR dj := 0 to 1 DO BEGIN
   temp := bndry(inp,k + 1,2*i + di,2*j + dj);
   store(outp,k + 1,2*i + di,2*j + dj,temp)
   IF temp > threshold THEN
   Edge-4-2(inp,outp,k + 1,2*i + di,2*j + dj)
   END;
   END;
   END;
END;
```

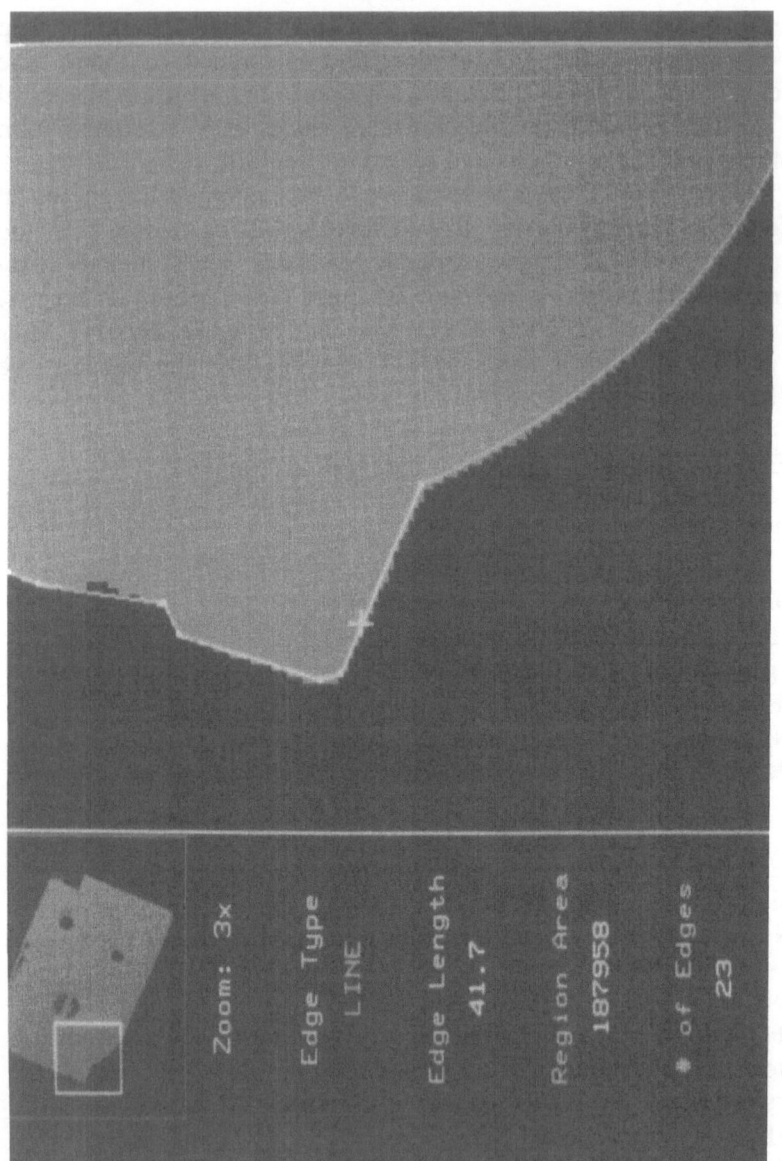

Figure 136. Line tracking by edge elements and line features. (Courtesy AdeptVision Inc.)

16.6. ALGORITHM LINE-1: EDGE TRACKING

The purpose here is to track, in firmware or software, a line in the edge image $e(i, j)$ by recursively trying to extend it in direction l by adding edge elements in direction l selected by an evaluation function $f(k)$ calculated for all neighbors k of (i, j) in a window W. The steps are as follows, and an example is given in Figure 136:

Step 1: Calculate for all k in W:

$$f(k) = |e(k)|^2 \cdot \cos[d(k) - d(i, j)]$$

where $|e(k)|$ is the edge intensity at neighbor point k to (i, j) in W, $d(k)$ is the angle of the edge at neighbor point k, and $K > 0$ is a constant.

Step 2: Select the direction l in which to extend the edge element $e(i, j)$ of angle $d(i, j)$:

$$l = \left\{ \underset{k \in W}{\operatorname{Max}} f(k) \text{ and } |e(k)| > K \right\}$$

Step 3: Move from (i, j) to $l \in W$ and replace (i, j) by pixel position at l. Return to step 1.

Line attributes such as the line length and the intensity of the pixels around the line can be collected at the same time.

Chapter 17

Region Segmentation and Boundaries

17.1. INTRODUCTION

At the medium image understanding level, it is often necessary to extract the objects in the image: this problem is called image segmentation. It can be approached by detecting edges and lines or by segmenting the image into distinct regions. Region segmentation focuses on the homogeneity of the image features, using the assumption that these features will be homogeneous or change gradually within the region corresponding to each object, but will change abruptly at the boundaries between objects. This approach can be viewed as a way of clustering pixels, and the various methods can be classified by the domain in which the clustering is performed: the image domain or the feature domain.

17.2. ALGORITHM SEGM-1: REGION GROWING

Region growing divides the image into a set of connected regions by clustering pixels; the clustering is performed in the feature domain. Several methods can be used:

1. Region growing by merging, where statistical hypothesis testing (Kolmogorov test, smoothed difference hypothesis test) helps decide whether regions and their neighbors have similar feature distributions,[132] or whether a one-dimensional strip approximation is followed by merging those regions whose approximation coefficients are close enough.[133,134]

2. Region segmentation by decomposition into rectangular sub-regions until homogeneous regions are achieved.
3. Split-and-merge, starting with an intermediate segmentation and splitting or merging regions as necessary[135]; this approach is rather fast.
4. Region growing by linking pixels, where a graph is constructed whose vertices correspond to the pixels and edges for pairs of connected pixels satisfying some relation[136–138]; the regions are grown by finding the connected components of the graph.

17.3. ALGORITHM SEGM-2: FEATURE DOMAIN CLUSTERING

Feature domain clustering methods classify and cluster picture elements or subregions in the feature space; the result is then projected back into the image domain. This approach does not, in general, take the spatial connectivity of the pixels in the image domain into consideration. Thus, the regions are not (in general) connected. Postprocessing, such as component labeling, may therefore be necessary to obtain the connected regions. The approaches are:

1. Classification of pixels by supervised learning, for example, by a nearest-neighbor rule.[139]
2. Clustering of pixels in the feature domain, by k-means clustering.[140,141]

17.4. ALGORITHM SEGM-3: SPATIAL CLUSTERING

Spatial clustering uses some method to partition the image into small connected regions and then clusters them in the feature space. This technique is insensitive to noise in classifying the pixels. Knowledge about the region shapes can also be taken into account. Examples are given in Refs. 142–146.

17.5. ALGORITHM BOUND-1: DETECTION AND TRACKING OF BOUNDARIES

To describe the shape of a region, the border of a binary labeled image can be detected by edge and line detection, and the attributes of this border can be measured. When the region is a connected component, the border is made out of one or more closed curves. When the region is simply connected, one closed curve which surrounds it can be extracted. When there are holes, additional closed curves must be extracted for each hole.

A boundary can be represented in several ways:

- As a line pattern in a 2-D array.
- As a series of coordinates.
- As a chain code representation.

Boundary detection algorithms can be divided into parallel methods, which detect boundary points using a discrimination measure with respect to the background, and sequential methods, which track the boundary from a point on the border which is given or detected by scanning.

17.6. ALGORITHM BOUND-2: REGION BOUNDARY EXTRACTION

In a binary image with black regions, Algorithm Bound-2 determines the boundary points of these regions—both external boundary points and internal boundary points in holes.

First, the binary image is raster scanned from top to bottom and from left to right to find all left and right endpoints il and ir of black line segments. Each such black segment is called a node of the line adjacency graph. The array il is also used to mark nodes which have been visited, by changing their sign.

Next, the procedure BOUND-2, given below, is invoked, which itself calls the procedures UP and DOWN; BOUND-2 produces the (x, y) coordinates of points on a boundary, in a list BL, ordered counterclockwise for external boundaries and clockwise for internal ones. The special symbols ## and # denote the start of a region and of a hole, respectively.

Procedure BOUND-2
Steps:
    Search the line adjacency graph, top to bottom and left to right. If $il(j, i) > 0$ then do block 11;
    Begin block 11;
    1. Place (j, i) in a stack S and # in BL. (New region.)
    2. While S is not empty do block 21;
        Begin block 21;

1. Remove the top element of S: (j, i). Set new $= 1$.
2. While $il(j, i) > 0$ do block 22;
   Begin block 22;
   1. If (new $= 1$) add # to BL followed by the left endpoint of node (j, i); also set new $= 0$. (New contour: either external boundary or a hole.)
   2. Set down $= 1$.
   3. While down $= 1$ do procedure DOWN.
   4. Add the right endpoint of (j, i) to BL.
   5. While down $= 0$ do procedure UP.
   6. If $il(j, i) > 0$ add the left endpoint of (j, i) to BL.
   End block 22;
   End block 21;
   End block 11;

Procedure DOWN
Steps:
   If node (j, i) has no children set down $= 0$; else do block 32.
   Begin block 32;
   1. Find its leftmost child $(j + 1, k)$.
   2. Place all other children in S.
   3. If $(j + 1, k)$ has a parent to the left of (j, i) then do block 33.
      Begin block 33;
      1. Set $i = i - 1$.
      2. Set down $= 0$.
      End block 33;
      Else do block 34.
      Begin block 34;
      1. Set $j = j$ and $i = k$.
      2. Add the left endpoint of (j, i) to BL.
      3. Mark (j, i) by setting $il(j, i) = -il(j, i)$.
      End block 34;
   End block 32;
   End DOWN.

Procedure UP
Steps:
   If node (j, i) has no parents set down $= 1$; else do block 52.
   Begin block 52;
   1. Find its rightmost parent $(j - 1, k)$.

2. If $(j - 1, k)$ has a descendant to the right of (j, i) then do block 53.

    Begin block 53;
    1. Set $i = i + 1$.
    2. Set down $= 1$.
    End block 53;
  Else do block 54;
    Begin block 54;
    1. Set $j = j - 1$ and $i = k$.
    2. Add the right endpoint of (j, i) to BL.
    End block 54;
  End block 52;
End UP.

17.7. ALGORITHM AOI-1: AREA OF INTEREST (AOI) PROCESSING

The simplest of all segmentation procedures is to use CAD or drill tape layout files (Algorithms CAD-1, Drill-1) to generate a mesh of

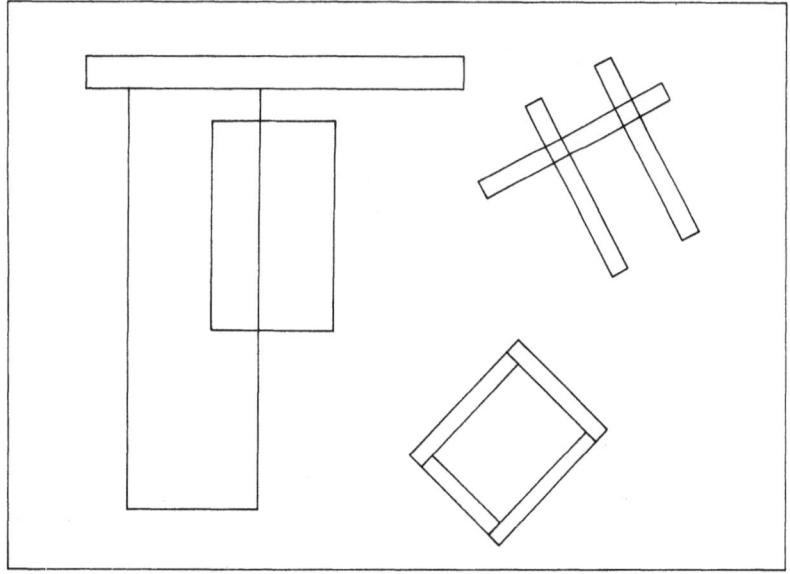

Figure 137. Area of interest processing (Algorithm AOI-1); the inclination should in general be inconsequential, although vertical and horizontal boundaries are easiest in view of the scanning.

typically rectangular regions (eventually overlapping) within the image, such that all further processing can be restricted to these rectangles individually or subsets thereof (see Figure 137).

This approach allows for local processing, for a reduction in the pixel data volume, and possibly for concurrent processing of the AOIs in parallel.

It should also be noted that low-resolution AOI imaging, for fixed AOI positions, can be achieved with fiber optic or LED detectors.

Higher-resolution AOI imaging can be achieved with, for example, split field microscopes, which show opposite corners of a scene at a larger magnification and with a higher resolution than is possible if the whole scene is viewed (as, for example, in TAB machine alignment).

Chapter 18

Geometry of Connected Components and Morphomathematics

18.1. INTRODUCTION

18.1.1. Goal

Methods are required for the determination and measurement of the geometrical features of regions or connected components. Operations for transforming these features are also needed. Since geometrical features describe the shapes of lines and connected components, they are usually defined from binary images, but extensions to gray-level images are also possible.

Since geometrical features are usually simple, and because they are used on a variety of levels in image analysis, many different algorithms exist which cannot easily be categorized. After some definitions, we will therefore proceed with reviewing the basic algorithms, without a categorization of them.

18.1.2. Adjacency (Connectedness) Relations

Only images represented by rectangular arrays are considered. Figure 138 depicts a 3 × 3 array. The set of points (1, 3, 5, 7) is called the set of 4-neighbors to the point 0, and the set of points (1, 2, 3, 4, 5, 6, 7, 8) is called the set of 8-neighbors to the point 0.

Based on the 4 (or 8) neighbors, 4 (or 8) adjacency is defined for pixels of binary value 1. When such 4 (or 8) adjacency is applied to the image, 8 (or 4) adjacency is applied to the background.

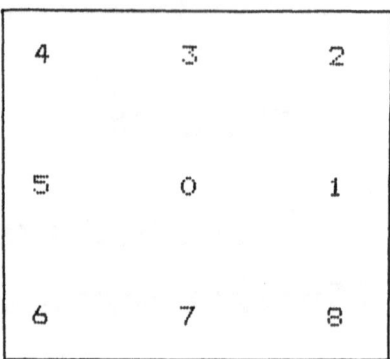

Figure 138. 3 × 3 pixel array.

18.1.3. Labeled Image

The labeled image is a two-dimensional array in which serial numbers have been assigned to each connected component (i.e., region). All pixels of the same region have the same label. Any two regions of the same image should have different labels.

18.1.4. Connected Components in Binary Images

Region segmentation divides a gray-level image into regions and produces a labeled image. An alternative approach is to label sequentially the connected components of a binary image and count the number of connected components obtained (Figure 139).

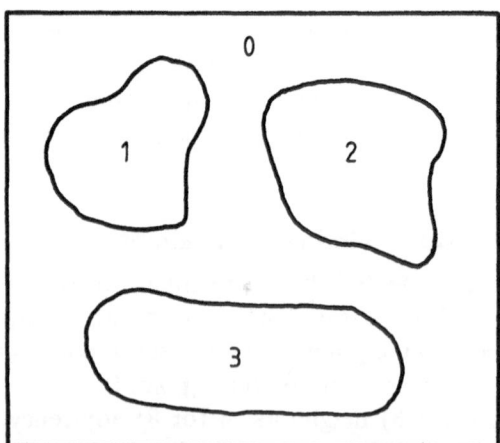

Figure 139. Connected components in a segmented (binary) image.

Whereas region segmentation generates regions which cover the image completely without gaps, connected component labeling labels only the components with pixels of binary value 1, leaving the points in the background. Thus, in connected component labeling, the background is treated differently than in region segmentation.

Connected components are sometimes also called "blobs."

18.2. ALGORITHM LABEL-1: LABELING OF CONNECTED COMPONENTS IN BINARY IMAGES

The procedure given below applies to binary images. It finds the connected components and labels them. If $x(i, j)$ is the input image, $y(i, j)$ is the same image with all pixels labeled, and the array $z(\cdot)$ gives the number and sequence of connected component labels. The steps are:

0. $N = 0$; set the dimension of $z(\cdot)$ at 1.5 to 3 times the estimated number of connected components.
1. Scan the input image vertically from the upper left to the lower right pixel. If the pixel value $x(i, j) < t$, where t is the gray-level threshold, set $y(i, j) = 0$; otherwise, check the labels $l(k, h)$ of the neighbors $[(i - 1, j), (i, j - 1)]$ for 2-connectedness and $[(i - 1, j - 1), (i - 1, j), (i - 1, j + 1), (i, j - 1)]$ for 8-connectedness.
2. If the labels $l(k, h)$ of these neighbors are all zero, set $y(i, j) = m$, and $z(m) = m$, assuming that $1, \ldots, (m - 1)$ have already been used as labels. Go to next point, $x(i, j + 1)$, and repeat step 1.
3. If the nonzero neighbors are all labeled by m, set $y(i, j) = m$, go to step 4, and repeat step 1.
4. If two or more different labels $n, o, \ldots$ are assigned to the neighbors, the labels must be changed to agree. Assign $m^* = \text{Min}(n, o, \ldots)$ to $y(i, j)$, and replace $z(m), z(n), \ldots$ by m^*. Replace $n, o, \ldots$ and so on in array $z(\cdot)$ with m^*. Go to step 5 and repeat step 1.
5. After scanning up to the last pixel in the image, change the numbers in the array $z(\cdot)$ to serial numbers, by changing the minimum value in $z(\cdot)$ to 1 and proceeding in ascending order. These serial numbers are the labels. Set the maximum label value at N.
6. Scan $y(i, j)$. If $y(i, j) > 0$, set $y(i, j) = z[y(i, j)]$.

18.3. ALGORITHM LABEL-2: RELAXATION LABELING

Whereas the labeling of connected components (Algorithm Label-1) assigns symbolic names (labels) to objects in the scene, some applications may require that such a labeling be consistent with a "world model" and that some uncertainties resulting from noisy, ambiguous image detail or large geometrical distortions be resolved.

The principles of relaxation labeling involve:

1. Defining a set of N possible disjoint labels (e.g., defect types).
2. Tentatively labeling the connected components based upon noisy, ambiguous image data; such an initial labeling can either be random or be provided by a human operator, the result of Algorithm Label-1, or the outcome of a comparison with, for example, a nominal CAD layout.
3. Iteratively adjusting the labeling to maximize the consistency with a world model (e.g., CAD) and to minimize labeling ambiguity.
4. Deferring the final labeling decision until no further progress is possible.

One example of relaxation labeling is given by Algorithm Edge-2, but Algorithm Label-2 is general. Other examples include region segmentation based on spatial continuity constraints (e.g., moving objects, stereo depth), vertex segmentation based upon geometrical constraints, or scene matching in general.

18.4. ALGORITHM LABEL-3: POSITION OF A CONNECTED COMPONENT

There are two methods for specifying the positions of connected components in a labeled image. One is to specify two points indicating the circumscribing rectangle of the component; the other is to specify one starting point on its inner border. When computing the features of the component, the computing cost for locating it in the image is reduced using the above position information.

A third approach, applicable in some special instances, is to use three-dimensional images from, for example, laser scanning, to position and align the image with respect to the height profile of the objects associated with the connected components. Such approaches are insensitive to variations in lighting and optical surface properties.

18.5. ALGORITHM MORPH-1: EROSION AND DILATION OPERATORS

Erosion and dilation are typical morphological neighborhood operations in images, both binary and gray level.[147] The elements are:

1. A square neighborhood, N.
2. The image $x(i, j)$.
3. A kernel or structuring element $b(i, j)$, containing, for example, 8-bit signed integers for gray-level processing or 1-bit logical values for binary processing.

The erosion e and dilation d of the image x by the structuring element b applied over the neighborhood N are defined by (Figure 140):

$$e(i, j) = \operatorname*{Min}_{l,m \in N} [x(i - l, j - m) - b(l, m)]$$

$$d(i, j) = \operatorname*{Max}_{l, m \in N} [x(i - l, j - m) + b(l, m)]$$

where the Min and Max are taken over the neighborhood, and where $(-)$, $(+)$ are gray-level operations or AND, OR for binary images. Examples of structuring elements b used for PCB design rule validation (Chapter 9) are given in Figure 141.

18.6. ALGORITHM MORPH-2: EXPANSION–CONTRACTION FLAW DETECTION

The principle of this nonreference procedure is described in Figure 142.[148] The results of expansion–contraction and contraction–expansion operations are compared to the original image or to the CAD layout, and the differences are flagged as defects.

The expansion and contraction operations are performed in turn by neighborhood operations (see Algorithm Buff-1) or in hardware (Figure 143). By controlling the ratio of expansions to contractions, thinning or region growing can be achieved.[76,147]

However, defect sizes vary widely from very small pinholes to pattern distortions extending over a considerable area, and the above method will miss defects whenever their dimensions are larger than the allowed linewidths.

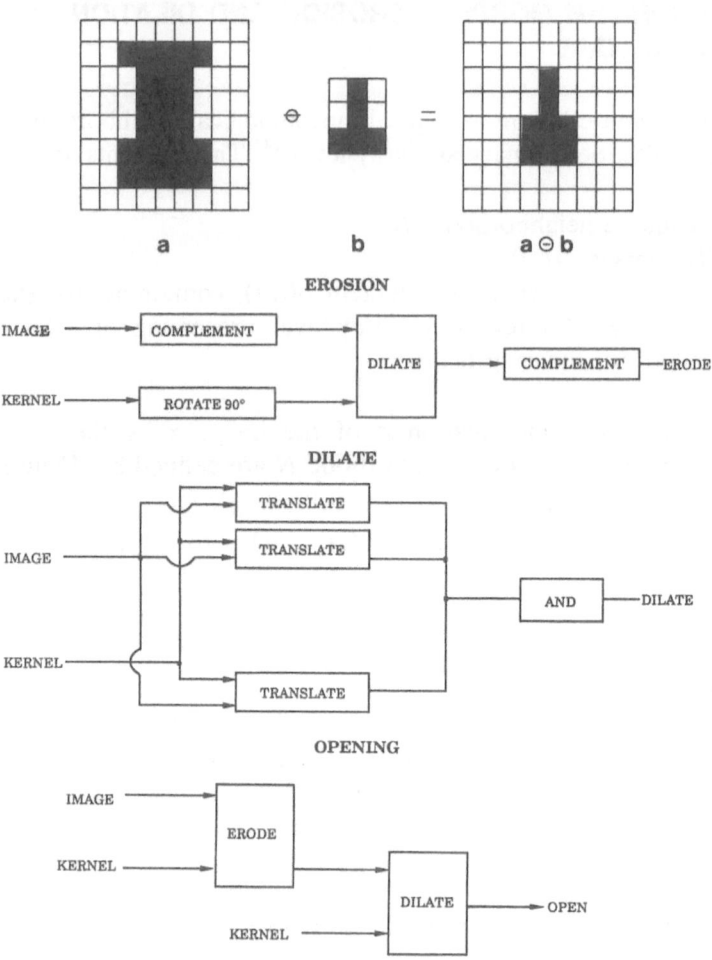

Figure 140. Erosion operation *e* of *a* by structuring element *b*.

The expansion and contraction operations are described as procedures GROW-OUT and TRACE-BACK in Tables 40 and 41 in the case of regions and in connection with a *routing* procedure. Figures 144 and 145 give examples of the results.[149]

The routing of a single wire, for example, takes three steps: first, GROW-OUT from the source *S* in the region a path of directions in *L* until target *T* is labeled in the region; second, TRACE-BACK along the source-pointing directions in *L* from the target to the source; last,

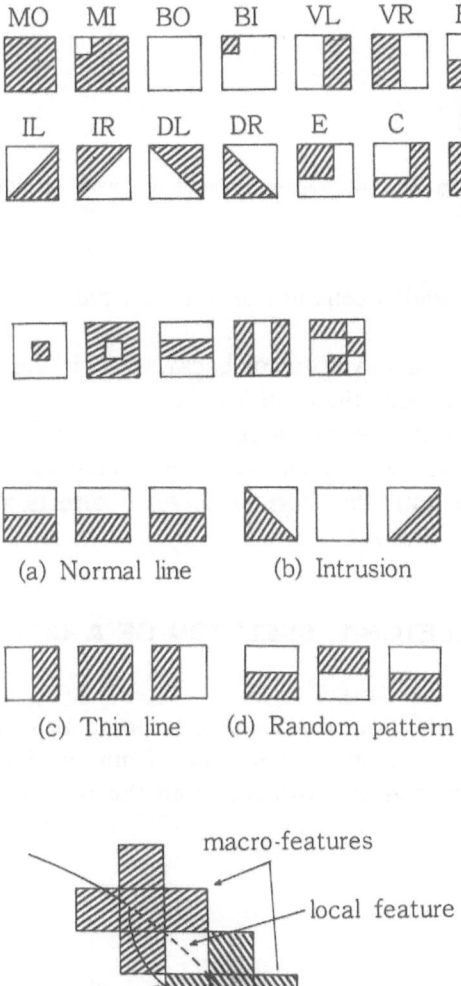

Figure 141. Structuring elements *b* for various PCB pattern defects.

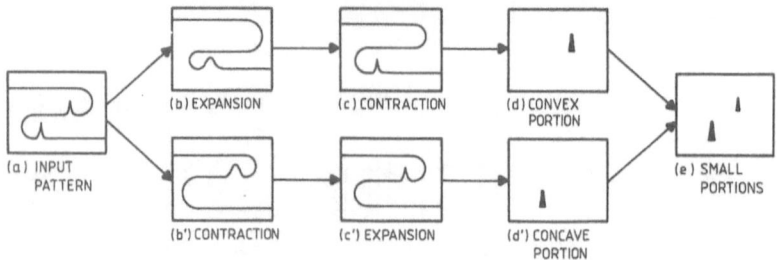

Figure 142. Parallel expansion and contraction operations, with resulting convex/concave/convex or concave defects.

CLEAN-UP extraneous labeled cells not on the new path and mark the new path as an obstacle.

GROW-OUT expands a wavefront of backward-pointing directions from the source S. When more than one direction may be chosen, the arbitrary order tests in the transform chooses directions in the order up, left, right, down. The operation "attach" appends directions.

GROW-OUT is run until the target receives a direction label, at which point TRACE-BACK is run.

18.7. ALGORITHM SKELETON-1: SKELETON OF A CONNECTED COMPONENT

The distance transformation replaces each point belonging to a connected component with its shortest distance from the background. Some geometrical features can be extracted from the resultant distance

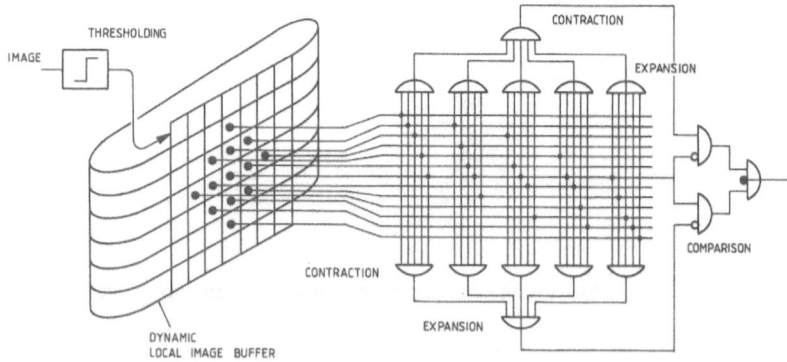

Figure 143. Hardware implementation of the expansion–contraction operation.

Table 40. GROW-OUT

The GROW-OUT, TRACE-BACK, and CLEAN-UP procedures are best thought of as neighborhood operations which change the value of the center pixel based on the result of set-membership tests on the pixels of the neighborhood.

Let

$$L = (\text{left, up, down, right, } S)$$

be called the Label set; membership in this set is tested in both the GROW-OUT and TRACE-BACK procedures.

GROW-OUT
IF center $\in$ (O, T)
THEN BEGIN:
  If n $\in$ L
  THEN attach up to center
  ELSE IF w $\in$ L
    THEN attach left to center
    ELSE IF e $\in$ L
      THEN attach right to center
      ELSE if s $\in$ L
      THEN attach down to center
END.

Above, S and T are source and target pixels. O and # are free and blocked cells. T followed by an element of L is a target labeled by a path pointing to source.

---

image. Of particular importance is the distance skeleton, consisting of the points within the connected component whose distance-transformed values are local maxima. The original image can be reconstructed from the skeleton points and the distances.[150] Different skeletons, however, are obtained for different distance functions; for example, the distance can be modulated by the sum of the gray levels on the local maxima path to the background.[151] Since the distance skeleton is defined as the

Table 41. TRACE-BACK

TRACE-BACK
IF center $\in$ L AND
  ((n = T . down) OR (w = T . right) OR (e = T . left) OR (s = T . up))
THEN attach T to center.

TRACE-BACK simply follows the directions from the target, attaching a T to each, until the source is reached and labeled ST. At this point, CLEAN-UP, a simple point-by-point transform, is run, which removes extraneous directions and labels the newly found path as an obstacle for future routes.

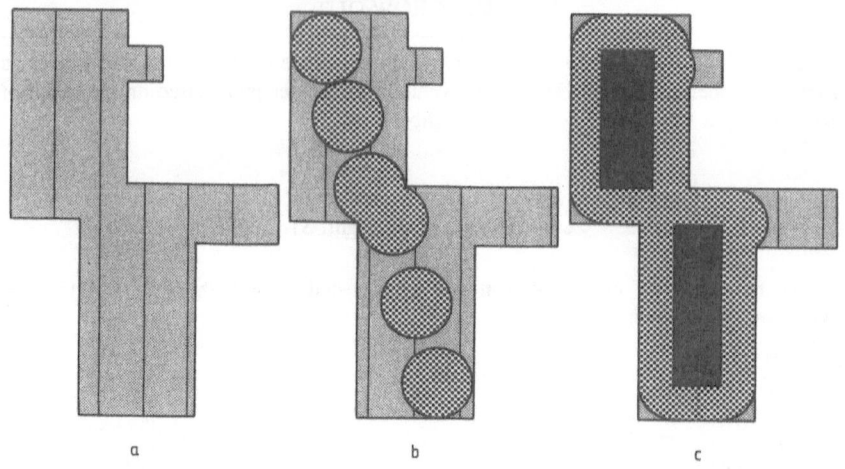

a   b   c

Figure 144. Result (c) of the expansion of (a) by sliding structuring elements in (b).[149]

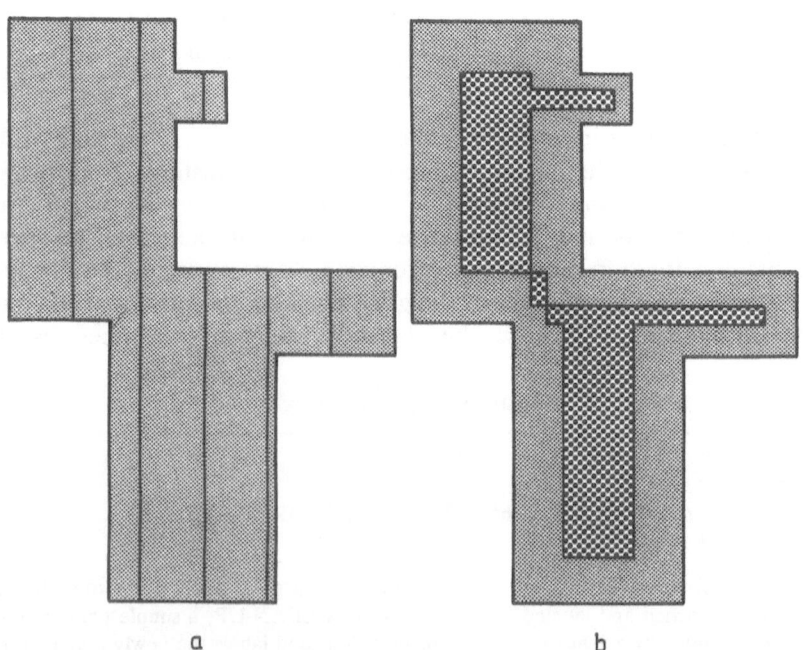

a   b

Figure 145. Result (b) of thinning of (a)[149]; see also Figure 143.

minimal set from which the region can be reconstructed, it does not always consist of connected line segments. The steps are:

A. Distance transformation of the input image $x(i, j)$ into the distance-transformed image $y(i, j)$ by a sequential iterative scanning (assuming 4-adjacency).

A1. Initialization:

$$Y^{(0)}(i, j) = \begin{cases} \text{large default value } M > 0, & \text{if } x(i, j) \geq 1 \\ 0, & \text{if } x(i, j) = 0 \end{cases}$$

A2. Iteration over $k = 1, 2, \ldots$ by:

$$y^{(k)}(i, j) = \text{Min}[Y^{(k-1)}(i, j); x(i, j) + y^{(k)}(i, j-1);$$
$$x(i, j) + y^{(k)}(i-1, j)]$$

$$Y^{(k)}(i, j) = \text{Min}[y^{(k)}(i, j); x(i, j) + Y^{(k)}(i, j+1);$$
$$x(i, j) + Y^{(k)}(i+1, j)]$$

where, for any given k, $y^{(k)}(i, j)$ scans in the forward direction from top to bottom and left to right, whereas $Y^{(k)}(i, j)$ scans in the reverse direction from bottom to top and right to left.

A3. Termination when $y^{(k)}(i, j) = Y^{(k-1)}(i, j)$, or $Y^{(k)}(i, j) = y^{(k)}(i, j)$, for all points (i, j), at which time $k = n$. When the input image is binary, the transformation ends after $n = 1$ iteration; for most gray-scale images, it ends after one or two iterations.

B. Skeleton extraction, where the skeleton is the set of points (i, j) satisfying the following conditions, with $l(i, j) = 1$ at a skeleton point and $l(i, j) = 0$ elsewhere:

$$y(i, j) + x(i, j+1) > y(i, j+1)$$
$$y(i, j) + x(i, j-1) > y(i, j-1)$$
$$y(i, j) + x(i-1, j) > y(i-1, j)$$
$$y(i, j) + x(i+1, j) > y(i+1, j)$$

This extracts the skeleton end point of the minimal path used to obtain the distance.

18.8. ALGORITHM SHRINK-1: SHRINKING OF CONNECTED COMPONENTS OR LINES

Shrinking is the process of successively contracting the connected components of a binary image until they are reduced to single points. It is used for counting connected components, lines, and other objects and making maps of their positions. The algorithm must make provisions to prevent components from disappearing after contraction to single points and must contain a condition for termination, which specifies the position and location of the final point.[152–154] Apart from these aspects, Shrink-1 uses the erosion e in Algorithm Morph-1.

18.9. ALGORITHM BRIDGE-1: MATCHING BRIDGES IN THE TOPOLOGICAL CELL LAYOUTS

Algorithm Bridge-1 is here exemplified in the case of IC cell images (see Section 5.2). The thinned preprocessed IC cell images obtained in Section 5.2 can be characterized as follows:

1. A graph G is obtained, the nodes (x_i) being either pad centers or angular points of the metallizations (and, as such, known through the topological CAD file); G is nonconnected like G^0.
2. This graph G approximates the graph G^0 (Figure 48), in which each point with abscissa s is labeled with, for example, the average thickness $\rho(s)$ of the metallization of the true IC in the window centered at this point. Other labels and the path algebra are given in Table 42.

It should be noted that for IC images, this thinning stage is far easier and computationally faster than any parsing of the IC etching boundaries. Smaller defects may, however, be eliminated, but they will then hopefully be picked up by Algorithm Fuzzy-1.

This Bridge-1 algorithm treats $G = (X, U)$ as an ordered, directed, labeled graph (see Table 42), characterized by the number, n, of nodes and an $(n \times n)$ matrix representation, $\mathbf{A} = (a_{ij})$, with entries:

$$a_{ij} = \begin{cases} l(x_i, x_j), & \text{if label of the arc } (x_i, x_j) \in U \\ \varnothing, & \text{if } (x_i, x_j) \notin U \end{cases}$$

where P is the path algebra (see Table 42), E is the neutral element of P, U is the order relation between nodes, $\varnothing$ is the null element of P, $l(x_i, x_j)$ is the label of the arc (x_i, x_j) in G, interpreted in Table 42; we will take

Table 42. Arc Labels $l(x_i, x_j)^a$ and Algebra $P,^b$ in G, with the Alternative Meanings of a_{ij}^* and $\hat{a}_{ij}$

Arc label $l(x_i, x_j)$ and algebra P	Significance of a_{ij}^* or $\hat{a}_{ij}$
0, 1 (binary pixels)	$a_{ij}^* = 1$ if x_j is accessible from x_i, and $a_{ij}^* = 0$ otherwise
$x \vee y = \text{Max}(x, y) \qquad x \cdot y = \text{Min}(x, y)$ $\varnothing = 0 \qquad\qquad\qquad E = 1$	$\hat{a}_{ij} = 1$ if there exists a nonnull path from x_i to x_j, and $\hat{a}_{ij} = 0$ otherwise
Physical length of the edge	a_{ij}^* is the length of the shortest path from x_i to x_j
$x \vee y = \text{Min}(x, y) \qquad x \cdot y = x + y$ $\varnothing = \infty \qquad\qquad\quad E = 0$	
Probability of existence of the edge	a_{ij}^* is the reliability of a most reliable path from x_i to x_j
$x \vee y = \text{Max}(x, y) \qquad x \cdot y = xy$ $\varnothing = 0 \qquad\qquad\qquad E = 1$	

^{a} x_i, y_j: nodes in the simple nondirected graph G; $l(x_i, x_j)$: label of edge (x_i, x_j) in G; $U = E$: order relation of this nondirected graph G. The edges (x_i, x_j) of the graph G are obtained by thinning of the electronic device image, and the values of the labels $l(x_i, x_j)$ are measured in the device subcell image. For example, l may be the gray level, i.e., emittance level related to temperature, of this arc. l may alternatively be the measured length of the thinned edge.
^{b} P has the operations $\vee$ (joint) and $\cdot$ (multiplication).

for $l(x_i, x_j)$ the average width or emissivity $\rho(s)$ over this arc (x_i, x_j) in the image represented by G (see Figure 48); $\vee$ is the joint operation in P (defined in Table 42), and $\cdot$ is the product operation in P (defined in Table 42) (with E as neutral element for $\cdot$).

A special label $l = 0$ is introduced for fictive edges to render the graph connected.

The idea of Algorithm Bridge-1 is not to match IC cell patterns but to match critical topological elements called bridges, which represent critical defects. The bridges are those sections of the IC cell surface which are nonredundant for proper electrical cell operations, e.g., open circuits and short circuits. Substrate interface anomalies also become parasite bridges. The graph theoretical definition of a bridge is illustrated in Figure 146[155,156]: in this figure, f is a bridge. The bridges are determined by the Algorithm Bridge-1, operating on the graph representation G of the cell image corresponding to the actual electrical excitation by some stimulus SE.

Let $H = (X, U)$ be the simple graph with the same nodes as G but with two arcs (x_i, x_j) and (x_j, x_i) between each pair of nodes x_i, x_j which are joined together by an edge in G; in H, the two arcs (x_i, x_j) and (x_j, x_i) both bear the name of the corresponding edge in G. An edge (x_i, x_j) of the simple graph G is then called a bridge of G, if in the graph

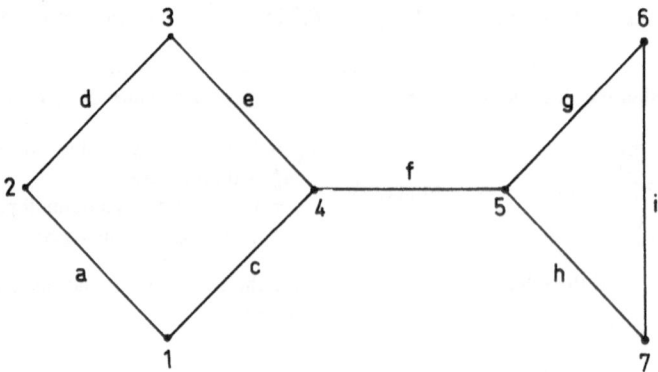

Figure 146. Example of a bridge in the subcell graph G derived from the subcell image. Here the bridge edge is f.

obtained from G by removing this edge, the nodes x_i and x_j are not connected (see Figure 146).

To compute these bridges, we must introduce $\mathbf{A}^{(k)}$, $\hat{\mathbf{A}}$, and $\mathbf{A}^*$. The kth power of $\mathbf{A}$, $\mathbf{A}^{(k)}$, can be defined in terms of the labels of paths on the graph corresponding to $\mathbf{A}$ in the following way. Let $S^{(k)}(i, j)$ be the set of all paths of order k from node x_i to node x_j on the labeled graph G of $\mathbf{A}$; then

$$a^{(k)}(i, j) = \bigvee \{l(s); s \in S^{(k)}(i, j)\}$$

Each element $a^{(k)}(i, j)$ of $\mathbf{A}^{(k)}$ is the set of names of all simple paths of order k from node x_i to node x_j.

We shall denote the strong and weak closure of a stable matrix $\mathbf{A}$ by $\mathbf{A}^* = [a^*(i, j)]$ and $\hat{\mathbf{A}} = [\hat{a}(i, j)]$, respectively. $\mathbf{A}^*$ is such that

$$\forall k = 0, 1, \ldots \qquad \mathbf{A}^{*(k)} = \mathbf{A}^{*(k+1)}$$

$$\mathbf{A}^* = E \vee \hat{\mathbf{A}} \qquad \mathbf{A}^* = \mathbf{A}^{(n-1)} \qquad \hat{\mathbf{A}} = \mathbf{A}\mathbf{A}^* = \bigvee_{k=1,\ldots} \{\mathbf{A}^{(k)}\}$$

Each element $a^*(i, j)$ of $\mathbf{A}^*$ is the set of names of all the simple paths from x_i to x_j. Each element $\hat{a}(i, j)$ of $\hat{\mathbf{A}}$ is the set of names of all nonnull simple paths from x_i to x_j. If binary labels only are considered (binary pictures), $\mathbf{A}$ is the Boolean adjacency matrix of the graph G; $\mathbf{A}^*$ has then entries $a^*(i, j) = 1$ if there exists any path from x_i to x_j, and $a^*(i, j) = 0$ otherwise, whereas $\hat{\mathbf{A}}$ has $\hat{a}(i, j) = 1$ if there exists any nonnull path from x_i to x_j, and $\hat{a}(i, j) = 0$ otherwise (see Table 42).

One simple algorithm to compute the bridges of G, that is, the open circuits and excessive short circuits in the IC cell, is then as follows.

Considering H, each entry of the closure $a^*(i, j)$ of its adjacency matrix $\mathbf{A}$ is the set of names of all the bridges between x_i and x_j. Thus, to find these bridges, we need to be able to compute $\hat{\mathbf{A}}$ or $\mathbf{A}^*$ directly.

One such method is the Jordan elimination method, which can be applied to compute the weak closure $\hat{\mathbf{A}}$. $\hat{\mathbf{A}}$ is the least solution $\mathbf{Y}$ of the equation $\mathbf{Y} = a\mathbf{Y} \vee \mathbf{B}$, if we set $\mathbf{A}^{(0)} = \mathbf{B}^{(0)} = \mathbf{A}$, because then $\hat{\mathbf{A}} = \mathbf{B}^{(n)}$. The steps of the algorithm are the following:

$$\mathbf{A}^{(0)} = \mathbf{B}^{(0)} = \mathbf{A}$$

$$\mathbf{B}^{(k)} = \mathbf{Q}^{(k)*}\mathbf{B}^{(k-1)} \qquad k = 1, 2, \ldots, n$$

$$\hat{\mathbf{A}} = \mathbf{B}^{(n)}$$

$$\mathbf{Q}^{(k)*} = \begin{bmatrix} E & B_{12}^{(k-1)} & B_{22}^{(k-1)*} & \varnothing \\ \varnothing & B_{22}^{(k-1)*} & B_{22}^{(k-1)*} & \varnothing \\ \varnothing & B_{32}^{(k-1)} & B_{22}^{(k-1)*} & E \end{bmatrix}$$

where the B_{lm} blocks are made of elements b_{ij}, given by

$$b_{ij}^{(k)} = \begin{cases} b_{ik}^{(k-1)} \cdot (b_{kk}^{(k-1)*}) & \text{if } j = k \\ (b_{kk}^{(k-1)})^* \cdot b_{kj}^{(k-1)} & \text{if } i = k \\ b_{ij}^{(k-1)} \vee b_{ik}^{(k-1)} \cdot (b_{kk}^{(k-1)})^* \cdot b_{kj}^{(k-1)} & \text{if } i, j \neq k \end{cases}$$

The closure of an element is defined in a similar way to the closure of a matrix.

Example (see Figure 146): Here $ is the empty label, and $\varnothing$ the zero element in G:

$$\mathbf{A} = \begin{bmatrix} \$ & a & b & c & \$ & \$ & \$ \\ a & \$ & d & \$ & \$ & \$ & \$ \\ b & d & \$ & e & \$ & \$ & \$ \\ c & \$ & e & \$ & f & \$ & \$ \\ \$ & \$ & \$ & f & \$ & g & h \\ \$ & \$ & \$ & \$ & g & \$ & i \\ \$ & \$ & \$ & \$ & h & i & \$ \end{bmatrix} \qquad \mathbf{A}^* = \begin{bmatrix} \$ & \$ & \$ & \$ & f & f & f \\ \$ & \$ & \$ & \$ & f & f & f \\ \$ & \$ & \$ & \$ & f & f & f \\ \$ & \$ & \$ & \$ & f & f & f \\ f & f & f & f & \$ & \$ & \$ \\ f & f & f & f & \$ & \$ & \$ \\ f & f & f & f & \$ & \$ & \$ \end{bmatrix}$$

Thus, f is a bridge, as shown by $\mathbf{A}^*$.

Finally, the bridges computed in each cell are matched against those of the reference layout represented by G^0, which was derived from the

CAD files. In other words, the defects represented by bridges (open or short circuits) are not the result of the pixel matching of the IC images with the reference layouts, but of a topological analysis on the thinned graph representation of the IC image.

Next, the interobject configuration in the IC and the reference symbolic description are verified for consistency in the matched sets of shapes and elements. Also, a hierarchy of branch-and-bound enumeration techniques helps in finding the ideal match by evaluating only a few combinations. The result is a failure detection and possibly localization.

Chapter 19

Feature Extraction

19.1. INTRODUCTION

Feature selection is the set of operations whereby one selects a set of quantified attributes for the shape, contents, and texture of a given line, connected component, or region. That set of attributes may be the same or different for all components or regions in a given image. The calculation of the attributes is carried out by feature extraction procedures. All feature values of a given component or region are usually grouped into a vector (called feature vector), a string, or a tree, for further processing; these data structures may include geometry features as discussed in Sections 9.3, 9.4, and 20.2.

19.2. ALGORITHM FEATURE-1: ATTRIBUTES OF A CONNECTED COMPONENT

On each labeled region, one may wish to measure attributes or features; all are digital approximations to the usual definitions for continuous figures. For small regions, the errors can be large.

Features which can be computed on one region with no starting point, with one starting point on the boundary, or with the circumscribing rectangle are:

1. Area.
2. Center of gravity.
3. Perimeter, with or without holes in the region.

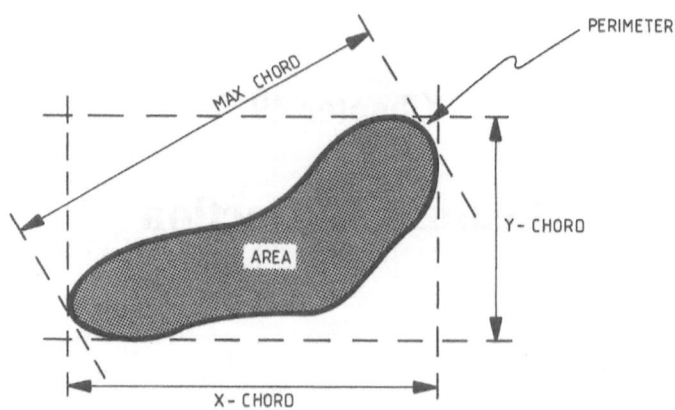

Figure 147. Chord, perimeter, and area of a connected component, with two derived features: p^2/A = perimeter^2/area; C/p = chord/perimeter.

EXAMPLES

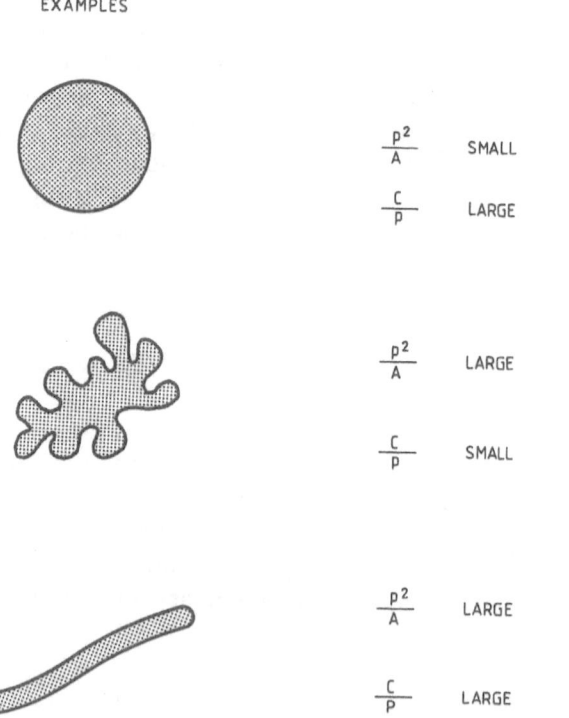

Figure 148. Typical values of the features in Figure 147 for different object shapes.

4. Size, with or without holes in the region (see Algorithm Label-3).
5. Compactness, with or without holes in the region.
6. Moments, with or without holes in the region (see Algorithm Moments-1).
7. Perimeter-to-area ratio, with or without holes in the region (Figures 147 and 148).

For all the above attributes, there are two computation approaches: either one can assume that the connected component is a binary one, with all pixels in it having the same label, or, once the pixels in the connected component have been found, one can reuse the true gray-level value of each pixel. For example, to calculate the area, one can (1) count the number of pixels in the connected component above some threshold or having the same label or (2) cumulate the gray-level pixel weights of all pixels in the connected component, and standardize these weights.

It appears that the second approach gives a higher repeatibility of the area measurement versus lateral displacements of the component over the spatial image quantization grid. The most repeatable measurements are those with small perimeter-to-area ratios, and those with edges which cut the pixel grid randomly so that individual pixel variations tend to cancel out. The least repeatable features are those about objects which have large perimeter-to-area ratios.

19.3. ALGORITHM FEATURE-2: HISTOGRAM FEATURES OF A REGION

Histogram features can be used to characterize AOI features by extracting the histogram curve in the AOI (normally in hardware) (see Algorithm Thresh-2) and calculating essentially moments of this curve (see Algorithm Moments-2). The histogram $f(n)$ gives in the AOI the number of pixels with gray level n over the range $0 < n < (2P)$ in a $(N \times N)$ AOI image array.

The basic steps are:

0. Select the AOI (Algorithm AOI-1), or carry out spatial masking with a lookup table (Algorithm LUT-1).
1. Extract the histogram $f(n)$ in the AOI (Algorithm Thresh-2), and eventually equalize it.
2. Calculate the features: Z_1, the height of the histogram peak, given by Max $f(n)$; Z_2, the width of the histogram peak, corresponding

to a 6-db window around Z_1; Z_3, the mean, given by

$$Z_3 = \frac{1}{N} \sum_{n=0}^{2P} n \cdot f(n)$$

Z_4, the standard deviation, given by

$$Z_4 = \sum_{n=0}^{2P} (n - Z_3) f(n)$$

Z_5, the energy, given by

$$Z_5 = \sum_{n=0}^{2P} [f(n)]^2$$

and Z_6, the entropy, given by

$$Z_6 = \sum_{n=0}^{2P} f(n) \cdot \log_2 f(n)$$

3. Recalculate the features from step 2 in histogram intervals, after partitioning the histogram range into sections associated with the presence of particular defects. These ranges will be placed absolute or relative to the histogram peak Z_1.

The classification routines using the histogram features above are dependent upon AOI size and location and upon defect types and sizes.

19.4. ALGORITHM MOMENTS-1: CALCULATION OF MOMENTS OF A CONNECTED COMPONENT OR LINE

Moments are a set of region attributes which can be computed digitally or on a hybrid processor[157–160] to characterize the shape of isolated binary regions. One class of moments are invariant to translation and rotation, while others are also scale invariant.

The moments of the region, with gray-level distribution $f(x, y)$, are:

$$m_{pq} = \iint x^p y^q f(x, y) \, dx \, dy; \quad p, q = 0, 1, 2, \ldots$$

The central moments are:

$$M_{pq} = \iint x^p y^q f[x + (m_{10}/m_{00}), y + (m_{01}/m_{00})] \, dx \, dy$$

$$M_{00} = m_{00}; \qquad M_{20} = m_{20} - (m_{10}^2/m_{00})$$

The invariant moments to be used as translation- and rotation-insensitive features are:

$$Z_1 = M_{20} + M_{02}$$

$$Z_2 = (M_{20} - M_{02})^2 + 4M_{11}^2$$

$$Z_3 = (M_{30} - 3M_{12})^2 + (3M_{21} - M_{03})^2$$

$$Z_4 = (M_{30} + M_{12})^2 + (M_{21} + M_{03})^2$$

$$Z_5 = (M_{30} - 3M_{12})(M_{30} + M_{12})[(M_{30} + M_{12})^2 - 3(M_{21} + M_{03})^2]$$
$$+ (3M_{21} - M_{03})(M_{21} + M_{03})[3(M_{30} + M_{12})^2 - (M_{21} + M_{03})^2]$$

$$Z_6 = (M_{20} - M_{02})[(M_{30} + M_{12})^2 - (M_{21} + M_{03})^2]$$
$$+ 4M_{11}(M_{30} + M_{12})(M_{21} + M_{03})$$

$$Z_7 = (3M_{21} - M_{03})(M_{30} + M_{12})[(M_{30} + M_{12})^2 - 3(M_{21} + M_{03})^2]$$
$$- (M_{30} - 3M_{12})(M_{21} + M_{03})[3(M_{30} + M_{12})^2 - (M_{21} + M_{03})^2]$$

Finally, invariant moments which are also scale insensitive are:

$$t_{pq} = M_{pq}/M_{00}^a$$
$$a = 1 + (p + q)/2; \qquad p, q: \text{integers}$$

19.5. ALGORITHM MOMENTS-2: SHAPE OF A LINE IN POLAR COORDINATES

The shape of a line is typically characterized by its polar coordinate representation, centered at a given fixed point, at several known fixed points, at the center of gravity, or at the border of a region to which the line is related. The line is scanned clockwise by vectors issued at the fixed points and extending to the line.

The representation includes the polar angles, θ_i, and polar distances, ρ_i, to points on the line, supplemented eventually by the slope or curvature at these locations on the line, with their moments calculated by Algorithm Moments-1.

An alternative approach consists in making a FFT expansion of (ρ, θ) to describe the shape of a closed curved line as a function of the curvilinear abscissa s along it[161]; when the line is specified using the slope as a function of the arc length s, the function is periodic, and the Fourier coefficients characterize the shape of the line.

19.6. ALGORITHM FUZZY-1: FIGURE OF MERIT FOR THE IC CELL FROM A FUZZY LANGUAGE DESCRIPTION

Algorithm Fuzzy-1 is exemplified by the IC cell case (Section 5.2). This algorithm is designed to compute for each IC cell, i.e., for each set of electrical stimuli SE, a figure of merit μ which accounts both for topological and geometrical faults. Lithographic resolution and image acquisition errors will have to be accounted for. It is better suited than Algorithm Bridge-1 for the detection of irregular growth/etching boundaries, scratches, blobs, and open circuits. Acceptance or rejection of the IC is on the basis of the figures of merit μ of all the cells.

Algorithm Fuzzy-1 relies on the following elements:

1. The topological model of the defect-free cell as a monoid V_0^* of a context-sensitive language L_0; V_0^* is generated by the CAD design software, restricted to the cell for which electrical testing is proceeding. V_0^* can be represented as in Figure 48.
2. A fuzzy class membership relation ρ for each string x of symbols generated in L_0, $0 < \rho(x) < 1$.[162] The actual value of $\rho(x)$ will be derived from the cell image as explained in Section 5.2, so as to enhance defects, while only taking into account defects whose sizes are in excess of process tolerances and optical resolution.
3. The recursive computation of the degree of agreement $\mu(x)$ of the actual cell x, with a fuzzy language[162] obtained by combining L_0 and the fuzzy relationship ρ; this degree of agreement is the figure of merit for the cell in the IC under test.

As explained in Section 5.2, the fuzzy class membership relation ρ is obtained from physical measurements in the IC cell image and from the design. For example, if a is a "primitive" etching shape element of L_0, the numerical value $\rho(a)$ can be a combination of the following criteria, which all are relevant IC quality features:

(a) Distribution of the spread of geometrical metallization width tolerances, including especially process tolerances (even if partially accounted for in the design rules).

(b) Area occupied by the shape a, e.g., a pad, as determined by thresholding the area and counting pixels. Any irregular offshoot, blobs, partial bridges, etc., then affect the values $\rho(a)$ for such pads.
(c) Weighting factor applied to $\rho(a)$, to account for the importance of that specific etching a.

It is essential to remember, in the remainder of this section, that the values of ρ are measured in the cell image according to some measurement definitions of which (a), (b), and (c) above are examples. The window thinning operator produces ρ as a by-product.

Let us define in detail the language L_0 of which the cell is a sentence. Let V_T be the finite set of primitives of the IC layout, including primitives associated with obvious geometrical defects. We denote by V_T^* the set of finite strings obtained by concatenation of primitives of V_T, including the null string $\emptyset$. The language L_0 is a subset of V_T^*, specified by the CAD design; it represents all acceptable layouts of the cell. The elements of L_0 may be generated by the grammar $G_0 = (V_N, V_T, P_0, S)$, where V_T is the set of terminals/alphabet, V_N is a set of nonterminals, $S \in V_n$ is the initial symbol, and P_0 is the finite set of production rules. The elements of P_0 are rewriting rules of the form $a \rightarrow b$, where a and b are strings in $(V_T \cup V_N)^*$. These rules are those by which the IC cell layout is obtained starting with $a = S$.

One important property of ICs, often overlooked in practice, is that the corresponding grammar G_0 is context sensitive for most circuit layouts: G_0 is said to be context sensitive iff the productions are of the form: $a_1 A a_2 \rightarrow a_1 B a_2$ with a_1, a_2, B in $(V_T \cup V_N)^*$, A in V_N, $B \neq \emptyset$, and $S \rightarrow \emptyset$ allowed.

We now define a fuzzy grammar[162] $\Gamma = (V_N, V_T, \pi, S)$ from G_0, by replacing P_0 by fuzzy rewriting rules defined as $a \rightarrow b\ (\rho)$, where ρ is the membership grade of string b, given a, or the figure of merit of b, given a, as obtained from the cell image (Section 5.2). Γ generates a fuzzy language L for which one can define the degree of properness $\mu(x)$ of any string $x \in V_T^*$, evaluating to which extent the string is correct with respect to G_0. More precisely, let $a_1, \ldots, a_m$ be strings in $(V_T \cup V_N)^*$ and $a_1 \rightarrow a_2\ (\rho_1), \ldots, a_{m-1} \rightarrow a_m\ (\rho_{m-1})$ be productions in π of Γ.

Then a_m is said to be derivable from a_1 in Γ. The string x of V_T^* representing the IC cell is said to be in L if x is derivable from S. The grade of membership $\mu(x)$ of x in L is then computed as follows:

$$\mu(x) = \sup \text{Min}[(\rho(S \rightarrow a_1), \ldots, \rho(a_m \rightarrow x)]$$

				a_{14}	a_{13}	a_{12}	a_{11}	a_{10}	a_9	a_8				
			a_{15}							a_7				
		a_{16}		b_{10}	b_9	b_8	b_7	b_6		a_6				
	a_{17}		b_{11}						b_5		a_5			
a_{18}			b_{12}		c_6	c_5	c_4		b_4			a_4		
a_{19}		b_{13}		c_7				c_3		b_3		a_3		
a_{20}		b_{14}	c_8					c_2		b_2		a_2		
a_{21}		b_{15}	c_9			d		c_1		b_1		a_1		
a_{22}		b_{16}	c_{10}					c_{16}		b_{28}		a_{40}		
a_{23}		b_{17}		c_{11}				c_{15}		b_{27}		a_{39}		
a_{24}			b_{18}		c_{12}	c_{13}	c_{14}		b_{26}			a_{38}		
	a_{25}		b_{19}						b_{25}		a_{37}			
		a_{26}		b_{20}	b_{21}	b_{22}	b_{23}	b_{24}		a_{36}				
			a_{27}							a_{35}				
				a_{28}	a_{29}	a_{30}	a_{31}	a_{32}	a_{33}	a_{34}				

$$
\begin{aligned}
M = \ &(a_1 \cdot a_{21} + a_2 \cdot a_{22} + a_3 \cdot a_{23} + \ldots + a_{20} \cdot a_{24}) \cdot \\
&(\overline{a}_1 \cdot \overline{a}_{21} + \overline{a}_2 \cdot \overline{a}_{22} + \overline{a}_3 \cdot \overline{a}_{23} + \ldots + \overline{a}_{20} \cdot \overline{a}_{24}) \cdot \\
&(b_1 \cdot b_{15} + b_2 \cdot b_{16} + b_3 \cdot b_{17} + \ldots + b_{14} \cdot b_{28}) \cdot \\
&(\overline{b}_1 \cdot \overline{b}_{15} + \overline{b}_2 \cdot \overline{b}_{16} + \overline{b}_3 \cdot \overline{b}_{17} + \ldots + \overline{b}_{14} \cdot \overline{b}_{28})
\end{aligned}
$$

$$
P = (c_1 \cdot c_2 \cdot c_3 \cdot \ldots \cdot c_{16} \cdot \overline{d}) + (\overline{c}_1 \cdot \overline{c}_2 \cdot \overline{c}_3 \cdot \ldots \cdot \overline{c}_{16} \cdot d)
$$

Figure 149. Binary neighborhood feature extraction: 15×15 mask is shown with logical mask values $a_1 - a_{40}$, $b_1 - b_{28}$, $c_1 - c_{16}$; Boolean cumulative functions M and P are given where a_i, b_j, c_k is the logical pixel value, $\overline{a}_i$, $\overline{b}_j$, $\overline{c}_k$ its logical complement, and "·" the logical AND. M gives the minimum distance error around the center pixel d, and P gives a hole or pinhole error at pixel d.

where the supremum is taken over all derivation chains from S to x. Consequently, the figure of merit $\mu(x)$ of subcell x is the degree of properness of the least proper link in the derivation chain generating the actual cell x, and $\mu(x)$ is calculated on the "best" chain.

G is said to be recursive if there is an algorithm which computes $\mu(x)$ recursively. Since G_0 is context sensitive, Γ is too. As it has been shown[163] that a fuzzy context-sensitive grammar is recursive, the figure of merit $\mu(x)$ of the cells x on the IC can be computed recursively.

In practice, this computation is simply the recursive computation of the supremum in the formula for $\mu(x)$ from the measured values $\rho(a_{i-1} \rightarrow a_i)$; the measurement sequence is triggered by the CAD file which describes the sequential scanning procedure.

19.7. ALGORITHM FEATURE-3: BINARY NEIGHBORHOOD FEATURES

In a given square neighborhood around a given known pixel location (i, j) in a binary image, one may want to calculate directional point-to-edge, edge-to-edge, or perimeter features, and to hardware implement this feature extraction. One approach is to apply a binary neighborhood mask around this location (i, j), accounting for border effects (Figures 133 and 149), and calculate Boolean or integer-weighted Boolean cumulative functions of the binary image pixel values, multiplied by the corresponding mask values, at selected point sets in the neighborhood. The values of the calculated functions then serve as features for later classification.

19.8. ALGORITHM TEXT-1: TEXTURE FEATURES

Texture is the property of a region in one image to display some approximately periodic features of similar shape, where the periodicity can apply to a combination of directions, sizes, colors, and object features. Whereas a human eye can often perceive texture, and discriminate among texture types, this capability is not easy to achieve through computer vision.

One approach[164] is as follows:

1. Calculate the co-occurrence matrix $P(l, m)$ across the region, where the co-occurrence probability, $0 < P(l + 1, m + 1) < 1$,

0	0	1	1
0	0	1	1
0	2	2	2
2	2	3	3

$$P = \begin{bmatrix} 4 & 2 & 1 & 0 \\ 2 & 4 & 0 & 0 \\ 1 & 0 & 6 & 1 \\ 0 & 0 & 1 & 2 \end{bmatrix} \qquad P = \begin{bmatrix} 6 & 0 & 2 & 0 \\ 0 & 4 & 2 & 0 \\ 2 & 2 & 2 & 2 \\ 0 & 0 & 2 & 0 \end{bmatrix}$$

(a)   (dj, di) = (1,0)   (b)   (dj, di) = (0,1)

$$P = \begin{bmatrix} 2 & 1 & 3 & 0 \\ 1 & 2 & 1 & 0 \\ 3 & 1 & 0 & 2 \\ 0 & 0 & 2 & 0 \end{bmatrix} \qquad P = \begin{bmatrix} 4 & 1 & 0 & 0 \\ 1 & 2 & 2 & 0 \\ 0 & 2 & 4 & 1 \\ 0 & 0 & 1 & 0 \end{bmatrix}$$

(c)   (dj, di) = (1,1)   (d)   (dj, di) =(1,1)

Figure 150. Co-occurrence matrices (a)–(d) for the 4 × 4 × 4 bit top image array; (*di*) is a row change, and (*dj*) a column change.

is the frequency by which a pixel x^*, separated by a spatial displacement ($di = -1/0/1$, $dj = -1/0/1$) from pixel x having gray level l, will itself have gray level m; that frequency is estimated for x^* over the whole region (Figure 150).

2. Characterize the texture type by features F calculated from the co-occurrence matrix $P(l, m)$ for a region with N gray levels. Such features include:

   • Contrast:

$$F_1 = \sum_{k=0}^{N-1} k^2 \left(\sum_{l=1}^{N} \sum_{m=1}^{N} P(l, m) \right) \Bigg|_{|l-m|=k}$$

- Correlation:

$$F_2 = \sum_{l=1}^{N} \sum_{m=1}^{N} \frac{((l \cdot m)P(l, m) - Q(l, \cdot)Q(\cdot, m))}{\sigma_l \cdot \sigma_m}$$

where $Q(l, \cdot)$ and $Q(\cdot, m)$ are the averages of the marginals of $P(l, m)$, and σ_l and σ_m are the corresponding standard deviations:

$$Q(l, \cdot) = \frac{1}{N} \sum_{m} P(l, m) \qquad Q(\cdot, m) = \frac{1}{N} \sum_{l} P(l, m)$$

- Entropy:

$$F_3 = -\sum_{l=1}^{N} \sum_{m=1}^{N} P(l, m)[\log P(l, m)]$$

19.9. ALGORITHM SHADE-1: SHADE FEATURES

Shade features tell something about the presence or absence and about the orientation of the objects which generate the shade. Within each region containing binary shade with shade quantization 0, the

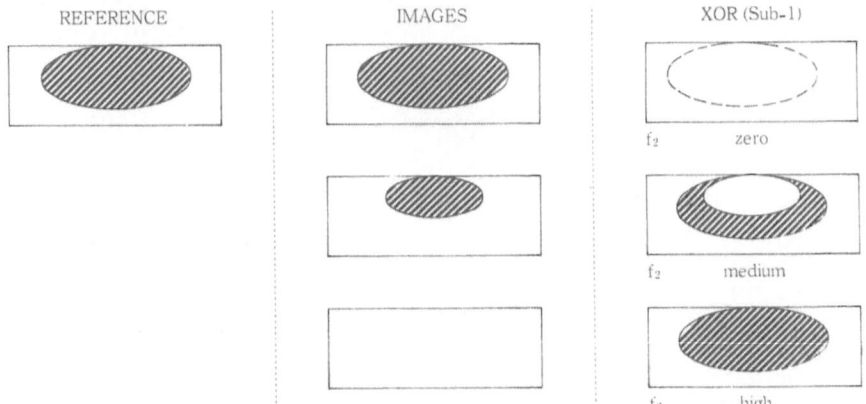

Figure 151. Shade features.

following features are computed (Figure 151):

F_1: Minimum of the line pixel averages, among all lines in the region; F_1 is equal to 1 if there is no shade, and 0 if the region is all shade.

F_2: Integer count of the number of pixels which are different between the region image and the reference image of the same region with nominal shade; F_2 is small if all is in order.

Missing components correspond to F_1 high and F_2 medium to high; improperly mounted components correspond to F_1 high and F_2 low to medium.

Chapter 20

Decision Logic

Essentially two basic decision logic approaches will be briefly reviewed: they serve to classify defect types on the basis of vision-derived features (see Chapter 19).

20.1. ALGORITHM CLASS-1: PATTERN CLASSIFICATION

The first approach encompasses all classification procedures from statistical pattern recognition[107] as well as the newer neural net classifiers.

The following three-stage process is applied:

1. Determine a catalog of $(N + 1)$ defect classes $E_0, E_1, \ldots, E_N$, including the no-defect class E_0.
2. Derive feature vectors $\mathbf{F} = (F_i)$ and classification rule parameters from sets E_j^* of training instances of all classes $j = 0, \ldots, N$.
3. Apply the classification rules found in step 2 to any new unsupervised instance $\mathbf{F}$ to be classified.

The classification rules are functions d_j defined by:

$$d_j(\mathbf{F}) = E_j, \qquad j = 0, \ldots, N$$

Because of the training step (step 2) and finite training sets, only estimated procedures d_j^* are known, and classification errors result from the differences between the true classification outcome $d_j(\mathbf{F}) = E_j$ for $\mathbf{F}$ truly of class E_j and the estimated classification outcome $d_j^*(\mathbf{F}) = E_j^*$ for $\mathbf{F}$ truly of class E_j.

The simplest classification rules are:

1. The minimum distance classifiers, which, given a metric $D(\cdot, E)$ between a feature vector and a training class represented by its samples, decide as follows:

$$d_j^*(\mathbf{F}) = E_j^* \quad \text{iff} \quad D(\mathbf{F}, E_j^*) = \operatorname*{Min}_i D(\mathbf{F}, E_i^*), \qquad i = 0, \ldots, N$$

   D can be, for example, the distance to the mean point in E.
2. The k-nearest-neighbor classifier,[107] where in $D(\cdot, E_i^*)$, E is reduced to the distance to the nearest sample in E_i^*.
3. The discriminant function classifiers, which, given a linear mapping $\mathbf{W}_j^* = (w_{jl}^*)$ trained from each class E_j^*, will decide as follows:

$$d_j^*(\mathbf{F}) = E_j^* \quad \text{iff} \quad \text{for all } i = 0, \ldots, N; \qquad i \neq j$$
$$\langle \mathbf{W}_i^*, \mathbf{F} \rangle = \sum_l w_{il}^* \cdot \mathbf{F}_l > 0$$

   where $\langle \ , \ \rangle$ is the scalar product.
4. The decision trees or semantic nets (Figure 152), which are sequences $s \in S_j$ of comparisons between linear mapping values

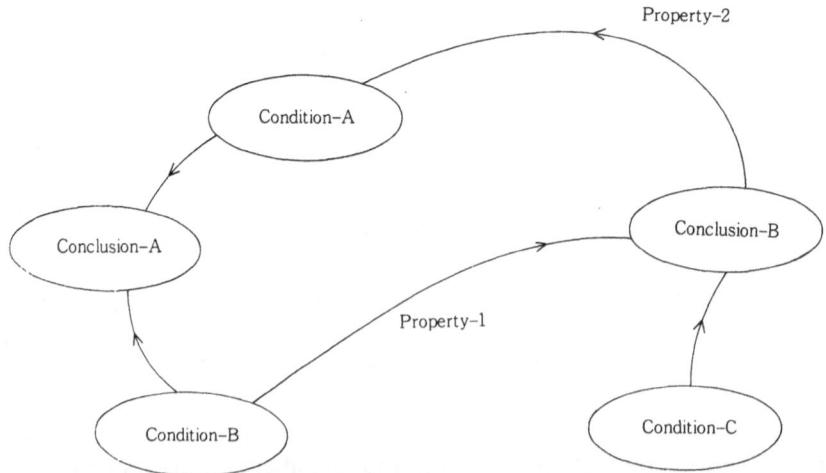

Figure 152. Semantic net for inferencing.

$\langle W_s^*, \mathbf{F} \rangle$ and fixed thresholds, with property-based branching conditions attached to each such comparison test outcome; ultimately, to each class E_j^*, $j = 0, \ldots, N$, should be attached a unique sequence S_j; each path S_j in the decision tree corresponds to a unique classification $d_j^*(\mathbf{F})$. See Section 7.2 for an example.

20.2. ALGORITHM RULE-1: RULE-BASED CLASSIFICATION

The second approach to classification covers all knowledge-based inference procedures (see Section 8.2.3), among which the simplest corresponds to rule-based processing. The advantages of this approach are that it allows for qualitative image feature interpretation and thus for defect classification despite image degradation and overlap between defect classes.[76,95]

Such decision rules applicable to a defect feature vector $\mathbf{F} = (F_i)$ have the general format:

Reference shape: p
Observed shape: $\mathbf{F}$
RULE: IF condition $(\mathbf{p}, \mathbf{F})$ THEN condition $(\mathbf{q}, \mathbf{F})$
Condition $(\mathbf{p}, \mathbf{F})$:

AND/OR
- Hypothesized class E_j
- $\langle W_p, \mathbf{F} \rangle = \sum_i w_{pi} F_i \geq 0$
- Logical condition on pattern/shape $\mathbf{F}$ compared to reference pattern/shape $\mathbf{p}$.
- Merge the pattern/shape $\mathbf{F}$ with the pattern/shape $(\mathbf{p})$, or split it, or delete it.
- Generate higher-level/more global pattern/shape features in $\mathbf{F}$, for later processing.

Rules such as the above allow the expression of the position and process dependency of the defects, especially as different image regions or CAD patterns are susceptible to different types of defects. See also Ref. 74.

Chapter 21

Image Data Structures and Management

Efficient image processing relies to a great extent upon the choice of data structures and knowledge representations which allow for a minimization of data transfer overheads and for the search of pointer or attribute lengths.

21.1. ALGORITHM CHAIN-1: CHAIN CODE REPRESENTATICN OF A LINE

A chain code representation of a line (including the border of a region) consists of (see Figure 153) the coordinates of the starting point on the line and a series of codes indicating the slope of line segments (or links) between successive points on the line ordered by their curvilinear abscissa values along it.

A line can be represented by chain codes in a compact way, and some transformations such as rotation of the line can be performed simply on the chain codes.[165]

21.2. ALGORITHM BUFF-1: FILTERING OPERATIONS WITH CIRCULAR BUFFERING

Neighborhood operations such as filtering involving, for example, 4-connected pixels must be performed frequently, thus leading to the requirement for their high-speed implementation. One approach is

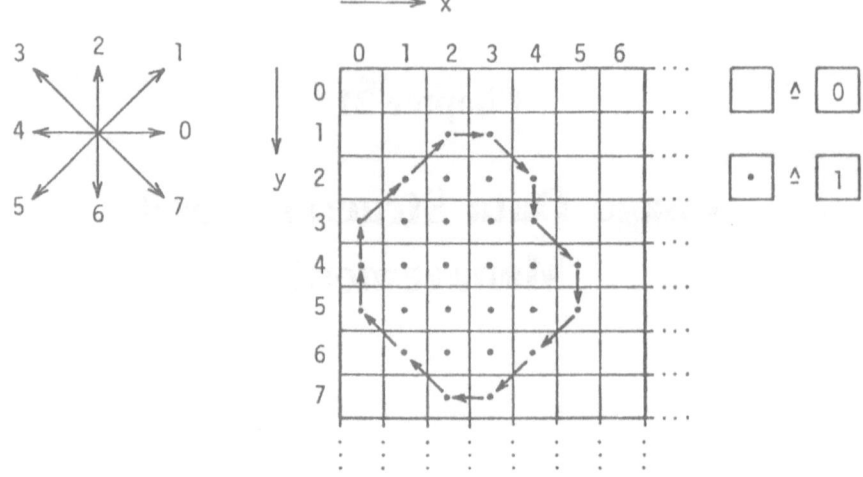

Figure 153. Chain code of order 8, i.e, with a slope quantization into eight possible values. The chain code with start at location (2, 1) is: $\{0, 7, 6, 7, 6, 5, 5, 4, 3, 3, 2, 2, 1, 1\}$.

circular buffering, described below for $3 \times N$ neighborhood operations and for possible parallel hardware implementation.

The idea behind circular buffering is to perform sliding operations in main memory, which delete raw image data as soon as possible (Figure 154). The circular buffering procedure is as follows:

1. Read image line $x(i - 1, \cdot)$ from peripheral memory, and store it in main memory starting at pointer address A; in parallel, read line $x(i, \cdot)$ and store from pointer address B; in parallel, read line $x(i + 1, \cdot)$ and store from pointer address C.
2. Compute the neighborhood operation, e.g., F filter (Algorithm Edge-1):

$$y(i, \cdot) = F[x(i - 1, \cdot), x(i, \cdot), x(i + 1, \cdot)]$$

3. Transfer $y(i, \cdot)$ to peripheral storage; read $x(i + 2, \cdot)$ from image peripheral storage, and store from pointer address A.
4. Compute $y(i + 1, \cdot)$, store the results from pointer address B, etc.

However, such a procedure leads to unknown values at the image borders in $y(i, j)$. One solution is in an initial step 0 to embed $x(i, j)$ into a larger array containing one layer of zeros all around and to discard the

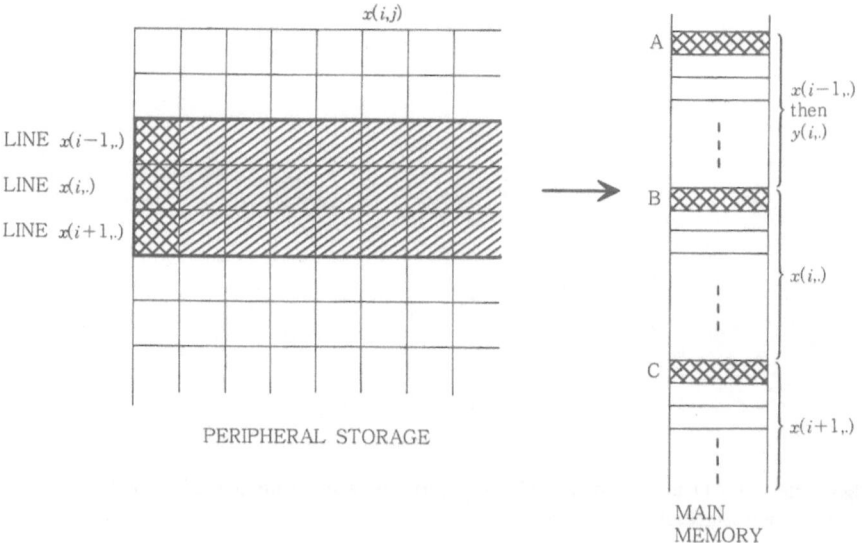

Figure 154. Circular buffering.

border of $y(i, j)$. An alternative is to augment $x(i, j)$ by replicating the border pixels on the outside layer, which should be one pixel deep.

21.3. ALGORITHM CAD-1: CAD DATA ORGANIZATION FOR IMAGE DATA COMPARISON

Several image processing and knowledge-based processing algorithms invoke computer-aided design or pattern generation layout files for reference (Figure 155). These layout files are here assumed to be organized into data structures of sequences of edges with location, connection, and jump conditions.

This CAD data structure is thus comparable with an image characterization from edge features and makes comparison of the image to the CAD layout possible.

When the image edge segments are compared to the CAD data base, the latter must be arranged in memory for rapid lookup. The tolerance data for each CAD edge may be included in the data base. This feature is extremely useful. A severe nick on a wide conductor cannot be detected by simple distance checking, and, on the other hand, a small narrow nick could cause a false alarm. The comparison method is able to see the difference between the absolute and the relative size of the defects by

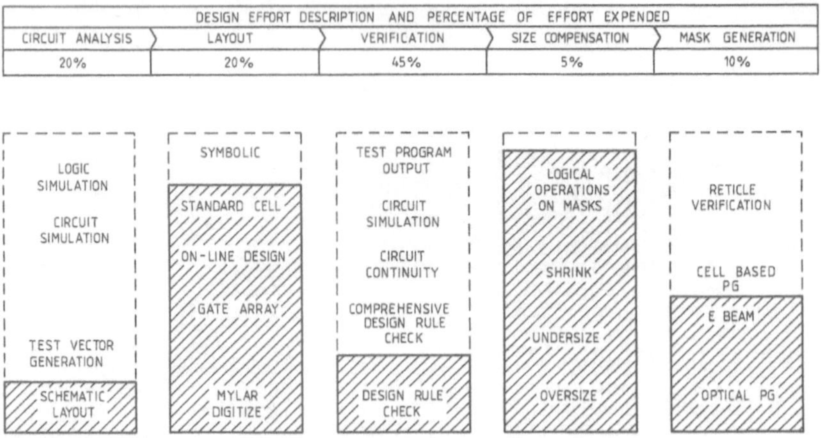

DESIGN EFFORT DESCRIPTION AND PERCENTAGE OF EFFORT EXPENDED				
CIRCUIT ANALYSIS >	LAYOUT >	VERIFICATION >	SIZE COMPENSATION >	MASK GENERATION
20%	20%	45%	5%	10%

Figure 155. CAD tasks and the relative efforts involved, with shaded areas highlighting functions performed already by commercial CAD systems.

flagging the edge segments extending or residing outside the tolerance limits.

The CAD data base is divided into pages, which are loaded into memory for the inspection of corresponding areas. Each page is, in turn, divided into subpages which are overlapped by the maximum allowed standard tolerance to avoid edge effects. The original CAD data file is modified for comparison by computing the edges of the details. Instead of storing both starting and ending points and possible radii of edges, we may compute intermediate results for the inspection stages. In fact, most of the computation can be done off-line.

More specifically, and when possible, the page/subpage data structure becomes a tree structure with attributes, each page being composed of a set of basic pattern commands and repeat commands. The basic pattern command directly specifies basic figures in each page and 2-D repetitions of subpages. Some algorithms normalize the description of the CAD pattern data and then automatically extract the repeated patterns and their arrangement.[66]

Each subpage is represented by a corresponding memory area. These areas are used for storing pointers to the edge descriptions. During comparison, the addresses of these pointer tables are determined from the starting and ending points of the line segments. We continue by extracting the common set of pointers from the tables. At this stage, we have usually only one candidate left. The descriptions are tried until the line segment fits the tolerance limits, or it is stored for error analyses.

The above data structures must be obtainable by conversion to or from standard CAD data formats (Calma, Applicon, PG, *e*-beam) or from pattern generation formats (Computervision, David Mann 3000, Electromask, etc.).

21.4. ALGORITHM DRILL-1: HOLE DRILL TAPE ORGANIZATION

In the case of bare boards with holes, the data structures to be used for the drill tape information are based on a sequence of hole-to-hole transition vectors $\mathbf{h}(dx, dy, dz, t, e)(n)$ where dx, dy, dz represent the three-dimensional spatial jump from hole center to hole center, t is the hole type (with pointers to the corresponding diameter and depth per type), and e is the tolerance on the hole center position. Repetitive transitions are replaced by the count (n) thereof.

21.5. ALGORITHM PAT-1: FILE AND VIDEO MANIPULATIONS OF GEOMETRICAL REFERENCE PATTERNS

The layout or other geometrical reference data files must be amenable to a number of digital or video transformations, which include

1. Scaling, both fine tuning, shrink, and expand, as well as magnification.
2. File merge, forming one data structure from two or more input files.
3. Windowing, that is, isolating a region by defining a window frame of interest.
4. Overlay removal, removing all areas of double exposure in specified overlays.
5. Tone inversion, by LUT reversal, producing a negative image from a positive definition and vice versa.
6. Proximity effect compensation, to compensate, for example, for inter- and intrapattern feature electron scattering effects.

Chapter 22

Conclusion: The Future of Computer Vision for Electronics Manufacturing

Three major trends in the future development of computer vision for electronics manufacturing can already be detected. First, vision and testing are the driving forces helping in identifying faults so that their causes can be successively eliminated, ultimately permitting the successive elimination of testers and, possibly, vision systems themselves. As a consequence, more and more feedback will be included into the process, through configuration, regulation, and explanation facilities.

Second, whereas it is still the exception, future vision systems will have much stronger ties to CAD, to repair systems, and to process data bases. From these sources, the vision systems will learn what to look for and what to ignore.

Third, just as mechanical probing is progressively being dislodged by testers, improvements in beam alignment or scanning speed, as well as in physical device characterization, will allow most low-level testers to be replaced by contactless probing (such as *e*-beam, ion beam, and photon excitation) associated with vision systems to interpret geometrical *as well as* electrical compliance with specifications. Gallium ion beams with beam diameters of 0.1 μm already allow for pattern geometries of 0.2 μm while eliminating photoresist altogether. In this way, contact resistance effects and most test sequence generation bottlenecks will be eliminated. This again represents a shift in emphasis from final testing to in-process probe testing.

In the shorter term, PCB design rule verification systems and IC inspection systems will have a higher commonality. At the same time, in the attached CAD systems, design for testability rules will be augmented with design for inspectability rules. Repetitive functions will increasingly be put into dedicated hardware, whereas inspection screens and vision system setup will be handled by knowledge-based processing.

Much care must still go into properly illuminating ever larger surfaces, while looking for ever smaller defects. Variations of angle, collimation, and intensity and differences in passivation or coating layers can produce different video signals, thus affecting inspection performance robustness.

Advances in computer vision itself will require more powerful segmentation techniques, guided by nominal geometrical patterns and incorporating methods of correcting for surface orientation effects. Structural descriptions must be hierarchically organized for more efficient matching, including fast elimination of inapplicable models. Benefits are of course also expected from research on concurrent image processing hardware architectures; accurate registration of multiple images for model building and image formation models; simple dyadic operators for Boolean operations on binary and gray-level images; recombination of imprecise information from many sources and relaxation processes for imprecise local evidence to yield globally consistent rules; calculation of statistical confidence values for all analysis, detection, classification, or understanding results; sensor fusion, especially for registration and feature extraction; operator interfaces; training and man/machine communication.

Appendixes

Appendixes

Appendix A

Glossary and Abbreviations

Alignment internally—Determines that each layer of the circuit is aligned (within some tolerance) to the previous layers.

Alignment with edge—Verifies that the circuit is aligned with the edge of the substrate (within some tolerance).

Alpha (α)—Usually refers to alpha particles from trace quantities of thorium and uranium in packaging materials. Alpha particles can cause soft errors (particularly in DRAMs).

Angstrom—1/10,000 micron.

AOI—Area of interest in an image.

Asher—A plasma-activated machine for removing photoresist. The operation is called ashing.

ASIC—Application-specific IC.

ATE—Automatic test equipment.

Backgate—A negative bias applied to an MOS gate through the channel region. See *Charge pump*.

BD—Brightfield/darkfield. A dual-mode technique used in reflected light microscopy. See *Brightfield* and *Darkfield*.

BI—See *Burn-in*.

BiCMOS—Bipolar CMOS process.

Bit—One unit of binary data. Either logical "1" or "0."

The reader is also referred to MIL-STD-1313 (Microelectronic Terms and Definitions) and Ref. 166.

Break—A broken geometry which can be repaired, caused by process problems or raw materials.

Bridge—A chrome link between two geometries. This can be removed by laser.

Brightfield—A normal operating mode in transmitted light work. In it, the illuminated object is visible against a lit background.

Bubbles—Usually refers to bubble memories. See *MBM*.

Burn-in—A general term for the process which exposes semiconductor or other devices to temperatures much higher than those encountered in use. Exposure time will vary depending on the temperature used, the kind of device being tested, and the purpose of the test. Strictly speaking, the term "burn-in" refers to a form of test designed to weed out infantile failures, contrasting with "life-aging," which is a test designed to provide quality information regarding specific devices. Typical burn-in temperatures range from 125°C to 200°C, although 300°C tests have been used for missile and space applications. The high-temperature environment is usually achieved by convected hot air in a closed chamber (or oven), but some power devices are tested by mounting them on high-thermal-inertia sinks through which a heated fluid is circulated.

Burn-in, dynamic—A high-temperature test conducted with the devices under test subjected to simulated actual operating conditions. The test simulation is commonly termed *exercising*, as input data signals and/or operating voltages are continuously varied during the test.

Burst refresh—Entire memory is refreshed before the next data access.

Byte—8 bits.

CAD—Computer-aided design.

CAS—Column address select (or strobe). Latches the column addresses into a multiplexed memory.

CAT—Computer-aided testing.

CCD—Charge-coupled device.

C-DIP—The standard ceramic package of the industry, employing side-brazed leads. See *CERDIP*.

CERDIP—The "economy" ceramic package of the industry, using a frit-sealed process.

Charge pump—A circuit (typically on a DRAM) to provide a backgate on devices by injecting electrons into the substrate. This backgate is necessary on NMOS devices to provide stability to threshold voltages. Also called substrate pump.

CMOS—Complementary MOS.

Controllability—A rough numerical measure of how easily the values of digital circuit nodes can be controlled from I/O pins.

Copolymer—Usually refers to a molding compound (mixture of silicone and epoxy) for plastic packaging. Example is Dow Corning's DC-631. Copolymers are much newer and not as well proven as Epoxy-B.

CVD—Chemical vapor deposition. Used extensively in the integrated circuits industry for applying thin films of materials to wafers (nitride/glass/etc.).

Darkfield—A transmitted light technique which ensures that any light not affected by the viewed object is kept out of the microscope eyepiece. The illuminated object is seen against a dark background, enhancing visibility.

Depth of focus (DOF)—The range over which a sharp image is given by a lens, at illuminating wavelength l, with numerical aperture N.A.: DOF = $l/(2 \cdot N.A.)$. For laser beams, see Chapter 7 in Ref. 167.

DFI—See *Dynamic fault imaging.*

DIC—See *Differential interference contrast.*

Dice—More than one die.

Differential interference contrast (DIC)—A transmitted light technique which is a powerful tool for analysis because it permits actual measurement of the refractive index and thickness of components of the viewed object. The technique produces brilliant color-contrast effects and reveals extremely fine details.

DIP—Dual-in-line component package.

Distributed refresh—Refreshing of the memory occurs between access cycles.

DLTS—Deep-level transient spectroscopy.

DRAM—Dynamic random access memory. See *RAM.*

Dry processing—The use of radio frequency sputtering and plasma reactants to etch and strip semiconductor wafers without the use of wet chemicals. See *Asher.*

DSW—Direct step on wafer, a step-and-repeat technique for exposing wafers. Gives better resolution than contact printing and increases mask usage. This is the exposure technique currently being used for most RAMs.

DUT—Device under test.

Dynamic fault imaging—Dynamic implementation of voltage contrast imaging.

e—Euler constant, $e = 2.17. \ldots$

EAROM—Electrically alterable ROM, similar to EEPROM but stores charge at a nitride–oxide interface.

e-Beam—Electron beam. Usually refers to the technique for mask making and wafer exposure using an electron beam as the energy source.

EBIC—Electron beam-induced current. See *SEM.*

ECC—Error code correction or error correcting code. Usually refers to a Hamming-type code which can correct single-bit errors and detect double-bit errors. Also called EDC.

EDC—Error detection and correction. See *ECC*.

EDS—Energy dispersive X-ray spectrometry.

EDX—Energy dispersive X-ray spectrometry.

EEPROM—Electrically erasable and programmable ROM. Similar to EAROM but stores charge on a floating gate. Newest types can erase individual data bytes.

Electromigration—A mass transport phenomenon resulting in a migration of metal in the direction of electron flow. Can result in open aluminum stripes and shorted junctions at high current densities.

EPI illuminator—A vertical illuminator with a beam splitter which directs light entering horizontally (from the source) vertically downward to the viewed object. Then, light is reflected up vertically from the viewed object through the beam splitter to the microscope eyepiece.

Epoxy-B—A Novolac-type of molding compound which is the standard plastic package for the industry.

EPROM—Erasable/programmable ROM. Almost always refers to the UV-erasable variety. See *UV*.

ES—Expert system. See also *KBS*.

ESCA—Electron spectroscopy for chemical analysis.

ESD—Electrostatic discharge. The primary cause of failure of input junctions and gates on MOS devices.

FAMOS—Field avalanche MOS. Refers to the technique for injecting charge onto the floating gate of UV EPROMs. Hot electrons are injected into the gate oxide by avalanching the drain junction into the channel.

Fault coverage—The percentage of potential stuck faults in an IC, device, or board that are uncovered by a set of test vectors; it is usually obtained through computer simulation.

Fault scanner—A device employed when a relatively large number of faults are to be summed for remote display, individually displayed, or processed. A circuit continuously scans the state of the fault sensors and transmits the state information in serial form to the point of use, where it is decoded by a demultiplexer. The demultiplexer restores the information to parallel form for the using circuit. The summed output may be used to activate the appropriate lamp and audible alarm on the power control and alarm panel, if desired.

Fault sensing—Faults due to presence or absence of voltage or current, whether steady state or pulsed, are sensed by comparators and compared with stable references. If a difference is sensed, a fault signal is emitted by the particular comparator. In the case of pulses, the comparator is clocked (or enabled) by a synchronous timing signal, and the comparator output is often provided with a latch circuit. The fault latches may be automatically or manually reset, depending on the use of the fault data. A mode switch is usually provided.

FFT—Fast Fourier transform.

FIT—Originally stood for failures in time; currently means a failure rate of one failure in one billion hours. Extensively used in Europe and Japan. More commonly used in the U.S. is percent per thousand hours or failures per million.

Flaky—Operating in an abnormal and unpredictable manner.

4" wafers—Actually 100 mm but 4" is close enough for "government work."

Front-end—Refers to a semiconductor wafer fab facility.

GaAs—Gallium arsenide.

Glass damage—Glass chips, scratches, or seeds caused by process problems or mishandling can cause printing errors.

Hidden refresh—Refreshing of the memory is done by the memory itself and is transparent to the user.

HMOS—One form of NMOS technology. Several vendors use this as their designation for a small-geometry/Si-gate/high-density NMOS process.

Hole detection—Detection of all holes (voids) in the printed material larger than some minimum size.

Hot electron—A highly energetic electron. Usually refers to an energy level sufficient to overcome the work function at a Si–SiO_2 interface, causing electron injection into a gate region.

Intrusion—The opposite of an extension; seen as a nick in a geometry. This can be repaired with a "hole plugger."

IR—Infrared (wavelength beyond $0.9\ \mu$m) radiation.

KB—Knowledge base.

KB/Kb—Kilobyte/kilobit.

KBS—Knowledge-based system.

KR—Knowledge representation.

l—Length of MOS channel. One of the most important physical parameters of scaled devices.

LIF—Laser-induced fluorescence.

LPCVD—Low-pressure CVD. See *CVD*.

LS—Usually refers to the low-power Schottky processs.

LSM—Laser scanning microscope.

LST—Laser scanning tomography.

MB/Mb—Megabyte/megabit.

MBM—Magnetic bubble memory. See *Bubbles*.

Micro—May refer to microprocessor or microcomputer.

Micron (μ)—One micrometer (μm). The basic unit of measure for VLSI geometries.

μP—Microprocessor.

Missing geometry—Any defect in this category will require the mask to be placed on hold for review by quality control engineering.

MNOS—Metal–nitride–oxide–semiconductor structure. Mainly used for gates in EAROM devices but was also the storage cell structure for the first 16-pin dynamic RAM.

Monitored BI—Dynamic burn-in while recording all read/write comparisons. Extensively used to test for alpha particles.

MONOS—Metal–oxide–nitride–oxide–semiconductor. A unique gate dielectric used by IBM in their RAM.

MOSFET—Metal–oxide–semiconductor field effect transistor.

MTBE—Mean time between errors. Refers to the failure rate of soft errors.

Mux—Short for multiplex.

N.A.—Numerical aperture.

Nibble—4 bits (half a byte).

Nitride—Usually refers to silicon nitride. Used extensively on VLSI devices for dielectrics/passivation/masking.

NMOS—*n*-Channel MOS.

Nonvolatile—A memory which retains data even after power-down, such as PROMs and MBMs.

Novolac—See *Epoxy-B*.

OBIC—Optical beam-induced current.

Observability—A rough numerical measure of how easily the values of digital circuit nodes can be determined from I/O pins.

OIF—Optically induced fluorescence.

Opaque spot—Due to their position, some opaque spots cannot be repaired. In this case, the piece must be scrapped.

Path comparison with standard—Comparison of the circuit pattern to the layout standard to detect syntactic mismatches.

Path spacing—Check that all paths in the image are separated by some minimum spacing.

Path width—Check that all paths in the image are wider than some minimum width.

Pattern generator—A circuit that generates a test pattern, usually for built-in testing; it may take any form, with random number generators and ROMs being the most common.

PCB—Printed circuit board.

%/KHrs—One of the standard failure rate designators used in U.S. Equals one failure per 1000 hours.

PG—Pattern generator format or software. Used with other CAD equipment to make a master mask reticle from digitized inputs.

PIND—Particle impact noise detection. Sometimes called PINT. The only quantifiable method of detecting loose particles (rattlers) inside packages.

Pinhole—Usually refers to dielectric failures caused by very small flaws at a photolithographic operation. Very often a latent defect that fails in the field. This type of defect will affect the area of the die if not repaired. Usually caused by either process problems in mask making or by the raw material used.

Pinspot—Small chrome spots which can normally be repaired by vaporizing the chrome with a laser. Usually caused by contamination during mask making.

Plague—Usually refers to one of the forms of Au–Al intermetallic growth. Purple plague is best known.

PLC—Programmable logic controller.

PLCC—Plastic leaded chip carriers.

PMAs—Precious metal adders. Used in many purchasing contracts to take into account the unpredictable price of gold on any particular day.

PMOS—*p*-Channel MOS.

PMT—Photomultiplier tube.

Poly—Usually refers to polycrystalline silicon. Extensively used with high doping as conductor/gate material. Also used in intrinsic state for high-valued resistors.

Polyimide—The most frequently used organic die coating to protect chips from alpha particles. Can be applied by several different processes.

Probe yield—Refers to the functional test of wafers just prior to being diced for assembly. Single most important yield point in a VLSI process. Typical VLSI probe yields are well under 40%.

Protrusion—An extension of chrome from the edge of a geometry which can often be removed with a laser, although not as critical as other defects. Usually the result of contamination during mask making.

PVS—Photovoltage spectroscopy.

PWB—Printed wiring board.

PZT—Piezoelectric transducer.

QC circles—A quality control technique developed in the U.S. but implemented in Japan. Forms workers into groups of (typically) eight people to develop better on-line work methods and techniques with emphasis on building-in quality.

RAM—Random access memory. Virtually all MOS memories are random access. RAM refers to a read/write chip.

RAS—Row address select (or strobe). Latches row addresses into a multiplexed memory.

Rattlers—See *PIND*.

RBEI—Robinson backscattered electron imaging.

RBS—Rutherford backscattering spectrometry.

Refresh—Recharging DRAM memory cells.

RMS—Root mean square.

s—Signal response.

SAM—Scanning Auger microscopy.

S/C—Abbreviation for semiconductor.

Scaling—The act of reducing geometries along the X–Y–Z axes together with adjustment of process parameters to effect a reduction in die size for a given circuit.

SE—Test stimuli voltage; also, in SEM context: secondary electrons.

SEAM—Scanning electron acoustic microscopy.

SECDED—Single error correct/double error detect. See *EDC*.

SEM—Scanning electron microscopy.

Sequential circuit—A digital circuit that change state according to an input signal (normally under clock control); it must be tested with a sequence of signals.

Shmoo plot—A series of plots showing the operating extremes of a device over a range of voltages or timing.

Shrink—Same as scaling.

SIM—Scanning ion microscopy.

SIMS—Secondary ion mass spectrometry.

SLAM—Scanning laser acoustic microscopy.

SMD—Surface-mounted device.

SMT—Surface mount technology.

Soft error—Refers to dropping or picking bits in a nonrepeatable manner. Often caused by alpha particles.

SOIC—Small-outline IC.

Solvent spot—Water and chemical stains, photoresist, fingerprints, etc. If contamination is excessive or is removable by ordinary cleaning procedures, the mask must be held for review.

SOT—Small outline transistor.

SRAM—Static RAM.

SSEM—Stroboscopic scanning electron microscope. A refinement of the SEM to allow noncontact probing of signals at various nodes of a chip.

Standby—Also called power-down. A mode whereby a volatile memory can retain data at a voltage level lower than normal operation.

STEM—Scanning transmission electron microscopy. See *SEM*.

Stuck fault—Usually a physical IC fault that results in one input or output of a logic gate improperly remaining either high or low regardless of the behavior of the circuits surrounding it.

Substrate pump—See *Charge pump*.

SV—Induced current.

TAB—Tape automated bonding component insertion machines.

Testability measure—A rough numerical indication of how easily test vectors can be generated for a particular circuit.

Test program—A computer program written in the language of a particular automatic production tester for ICs, PCBs, or devices.

Test vectors (test patterns)—A set of inputs and outputs generated for use in test programs.

T-Ox—Denotes oxide thickness. Usually refers to gate oxide thickness, which is one of the most important physical parameters of MOS devices. Gates on DRAMs are typically 300 angstroms.

TDDB—Time-dependent dielectric breakdown. Failure of gate oxides long after application of an electric field. This is the single most insidious failure mechanism of VLSI devices.

t-REF—Designates refresh time. One of the most important parameters of DRAMs.

Transmitted light technique—An illumination method in which light is transmitted through the viewed object.

Unknown—It is possible that automated inspection systems signal a defect where there is none, requiring careful visual inspection.

UUT—Unit under test.

UV—Ultraviolet (under 0.35 μm) radiation.

V-bump—Device failure caused by a level shift when a supply voltage is "bumped" even within the specified range.

VHSIC—Very-high-speed integrated circuit. Refers to the Department of Defense's VLSI Program.

VMOS—Vertical-groove MOS.

Voltage contrast imaging—The functioning of devices under normal operating conditions can be very graphically shown in several operating modes. This allows the tracing of failure propagation, the localization of malfunctions, and the interrogation of programs within the IC.

V_t—Threshold voltage. One of the most important electrical parameters of an

MOS device. VLSI chips typically have three to five different thresholds for thin-field and thick-field circuitry.

Wake-up cycles—Some dynamic circuits require a few pseudo cycles initially to pump boot strap nodes.

WD—See *Working distance*.

WDX—Wavelength dispersive X-ray spectrometry.

Working distance (WD)—The clearance between the objective lens and the viewed object (specimen). A $40\times$ objective might have a working distance of 0.53 mm. A $40\times$ extra-long-working-distance (ELWD) objective can have a WD of 10.1 mm.

XMR—X-ray microradiography.

XOR—Exclusive OR logic function.

XRF—X-ray fluorescence analysis.

$\times$—Optical magnification factor.

Appendix B

Relevant Journals

- *Artificial Intelligence in Engineering*
- *Assembly Automation*
- *Computer Vision*
- *Computer Vision, Graphics and Image Processing*
- *Electronic Materials*
- *Electronic Packaging and Production*
- *Electronics in Manufacturing,* Society for Manufacturing Engineers (SME)
- *Electronics Test*
- *European Semiconductor*
- *IEEE Design and Test Journal*
- *IEEE Expert Journal*
- *IEEE Transactions on Electron Devices* (Vol. **ED-**)
- *IEEE Transactions on Pattern Analysis and Machine Intelligence* (Vol. **PAMI-**)
- *Insulation Circuits*
- *International Journal for Hybrid Microelectronics*
- *International Journal of Pattern Recognition and Artificial Intelligence*
- *Microelectronic Engineering*
- *Microelectronics and Reliability*
- *Microelectronics Manufacturing and Testing*
- *Pattern Recognition*
- *Pattern Recognition Letters*
- *PC Fab*
- *Printed Circuit Fabrication*
- *Productronic*
- *Semiconductor International*
- *Solid State Technology*
- *Test and Measurement World*
- *Vision,* Machine Vision Association/Society for Manufacturing Engineers (MVA/SME)

Appendix C

Units and Conversion Tables

Common prefixes

Tera	T	10^{12}
Giga	G	10^9
Mega	M	10^6
kilo	k	10^3
milli	m	10^{-3}
micro	μ	10^{-6}
nano	n	10^{-9}
pico	p	10^{-12}
femto	f	10^{-15}

Albedo—Ratio of total radiation to total incident radiation, as a function of wavelength.

Candela (cd)—Unit of light intensity (see Figure 156).

$$10.764 \times \text{candle/ft}^2 = \text{candela/m}^2$$

$$3.4263 \times \text{foot-lambert} = \text{candela/m}^2$$

If the source is not a point but an extended surface, its luminance in cd/m^2 in a given direction is the luminous intensity per unit projected area of the normal to the direction considered, for a blackbody surface of $1/600{,}000 \, m^2$ in area, at the temperature of solidification of platinum. A joint source of one candela intensity radiates one lumen into a solid angle of one steradian.

FOOT LAMBERTS to candela/m² ft lamberts x 3.426 3 E+00= cd/m²

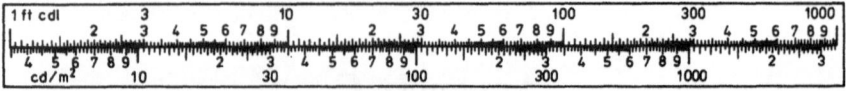

CANDLEPOWER to lumens cdl pr x 4π E+00 =lumens

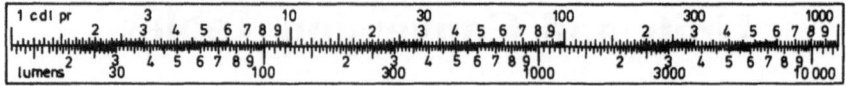

FOOTCANDLES to lumens/m² (lux) ft cdl x 1.076 4 E+01= lumens/m² (lux)

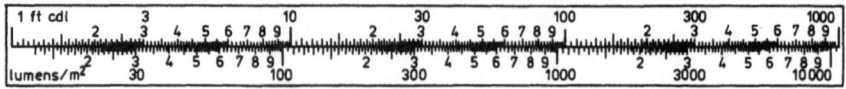

Figure 156. Conversion charts: foot lamberts to candela/m² (top), candlepower to lumens (middle), and footcandles to lumens/m²(lux) (bottom).

Degree per second, angular (°/s)—Unit of angular velocity.

$$\times\ 1.7453 \times 10^{-2} = \text{rad/s}$$
$$\times\ 0.1667 = \text{revolutions/min (rpm)}$$

Foot (ft)—Unit of length.

$$\times\ 0.3048 = \text{m}$$
$$\times\ 12 = \text{in}$$

Inch (in)—Unit of length.

$$\times\ 2.54 \times 10^{-2} = \text{m}$$
$$\times\ 8.3333 \times 10^{-2} = \text{ft}$$

Joules (J)—Unit of energy.

$$\times\ 2.778 \times 10^{-7} = \text{kw-h}$$
$$\times\ 2.390 \times 10^{-4} = \text{kcal}$$

Lambert (L or lb)—Unit of luminous flux (see Figure 156). One lambert is the luminance of $(1/\pi)$ cd/cm².

Lumen (lm)—Radiant flux W/unit wavelength.

$$\times\ (1/4\pi) = \text{candle}$$
$$F\,(\text{lm}) = c\int_{400}^{780} v(\lambda)P_\lambda\,d(\lambda)$$

where $c = 660$ in lm/W (nm), $v(\lambda)$ is eye response in lm/unit wavelength (nm), and P_λ is radiant flux in W/unit wavelength (nm). A point source of 1 cd emits (4π) lm into space.

Luminance (lm/sr $\cdot$ m²)—Luminance is given by

$$\partial^2\phi_v/\partial\omega \cdot \partial A \cdot \cos\theta$$

where symbols are as for radiance, and ϕ_v is the luminous flux.

Lux (lx)($=$ lm/m²)—Unit of illuminance (E).

$$1\, lx = 0.0929 \text{ footcandle}$$
$$\times\ (1/10{,}764) = \text{footcandle}$$
$$= \text{lm/ft}^2$$
$$\times\ 10^{-4} = \text{phot (lm/cm}^2)$$
$$E = I\cos\phi/d \qquad (\text{lm/m}^2)$$

where I is the equivalent point source luminous intensity in candela, ϕ is the angle between the line of sight and the line between source and receiver, and d is the distance between source and receiver.

Micrometer (μm)—Unit of length.

$$\times\ 1.000 \times 10^{-6} = \text{m}$$

Millimeter (mm)—Unit of length.

$$\times\ 39.4 = \text{mil}$$
$$\times\ 1000 = \text{micrometer } (\mu\text{m})$$
$$\times\ 3.94 \times 10^4 = \text{microinch } (\mu\text{inch})$$
$$\times\ 10^6 = \text{nanometer (nm)}$$
$$\times\ 10^7 = \text{angstrom (Å)}$$

Mil (mil)—Unit of length.

$$\times\ 10^{-3} = \text{inch}$$
$$\times\ 0.0254 = \text{millimeter (mm)}$$
$$\times\ 25.4 = \text{micrometer } (\mu\text{m})$$
$$\times\ 2.54 \times 10^5 = \text{angstrom (Å)}$$

Radian (rad)—Unit of plane angle.

$$\times\ 57.296 = \text{degree (angular)}$$

Radiance (W/sr · m^2)—Radiance is given by

$$\partial^2\phi/\partial\omega \cdot \partial A \cdot \cos\theta$$

where ∂A is a surface element, θ is the direction between the normal to A and the angle of observation, and $\partial\omega$ is the observation angle for $\partial\phi$.

Radiant emittance (W/m^2)—Radiant emittance is given by $\partial\phi/\partial A$ from area A.

Radiant flux (W)—1 W = $(1/h\nu)$ = 5×10^{18} photons/s.

Radiant intensity (W/sr)—Radiant intensity is given by $\partial\phi/\partial W$.

Reflectance—Ratio of reflected radiation to incident radiation, as a function of wavelength, angle of incidence, and angle of observation.

Second (")—Unit of angle.

$$\times \, 4.8481 \times 10^{-6} = \text{rad}$$
$$\times \, 2.778 \times 10^{-4} = \text{degree (angular)}$$

Square millimeter (mm^2)—Unit of area.

$$\times \, 1550 = \text{mil}^2$$

References

1. J. Tassone, An illumination system for inspecting PCB with surface mounted components, *Vision* **4**(4), 1–5 (1987).
2. G. E. Crook and B. G. Streetman, Laser based structure studies of Si and GaAs, *IEEE Circuits and Devices Magazine* **1986,** Jan., 25–31 (1986).
3. N. Khurana, Pulsed IR microscopy for debugging latch-up CMOS products, Proc. IEEE Int. Reliability Physics Symposium, Conf. 1984, pp. 122–125.
4. B. G. Cohen, Some applications to solid state devices of infrared microscopy, Technical Research Bulletin No. 1, Research Devices Inc., Berkeley Heights, N.J. (1982).
5. L. W. Kessler, Acoustic microscopy: A nondestructive tool for bond evaluation on TAB interconnections, Proceedings of the 1984 International Society for Hybrid Microelectronics (ISHM) Symposium, 17 Sept. 1984, Dallas, Tex.
6. C. R. Elliott, Advances in the infrared microscopy of electronic materials, *Proc. SPIE* **368** (1982).
7. V. South, Accuracy and precision in automatic video inspection systems, *Test and Measurement World* **1987,** May, 96–104 (1987).
8. J. Brakenhoff, Confocal scanning light microscopy with high aperture immersion lenses, *J. Microscopy,* **117,** Part 2, 219–232 (1979).
9. G. V. Lukianoff *et al.,* Electron beam testing of VLSI dynamic RAMs, *Electronics Test* **1982,** June, 46–56 (1982).
10. M. E. Levy, An investigation of flaws in complex CMOS devices by a scanning photoexcitation technique (laser), Proc. IEEE Int. Reliability Physics Symposium, 1977, pp. 44–53.
11. D. J. Burns and J. M. Kendall, Imaging latch-up sites in LSI CMOS with a laser photoscanner, Proc. IEEE Int. Reliability Physics Symposium, 1983, pp. 118–121.
12. T. C. May *et al.,* Dynamic fault imaging of VLSI random logic devices, Proc. IEEE Int. Reliability Physics Symposium, Las Vegas, Nev., 1984, pp. 222–228.
13. J. M. Patterson, Semiconductor junction temperature measurement using EBIC mode in SEM, Silicon Systems, Santa Clara, CA (1979).

14. V. Wilke, U. Goedecke, and P. Seidel, Laser scan microscope, C. Zeiss, D 7082 Oberkochen (1981).

15. R. Mueller, Scanning laser microscope for inspection of microelectronic devices, *Siemens Forsch. Entwicklungs Ber.* **13**(1), 9–14 (1984).

16. W. J. Alford and R. D. Vanderneut, Laser scanning microscope, *Proc. IEEE* **70**, 641–651 (1982).

17. E. Ziegler, Faktor analyse und Funktionspruefung von Halbleiterbauelementen, *Productronic* **1986**(4), 54–57 (1986).

18. H. M. Haskal and A. N. Rosen, *Appl. Opt.* **10**, 2775 (1971).

19. C. J. Sheppard, Applications of scanning optical microscopy, *Proc. SPIE* **368**, 88–95 (1982).

20. R. A. Smith, *Semiconductors*, Wiley, London (1984).

21. M. Schneider, Using near-IR microscopy for semiconductor production, *Microelectronics Manufacturing and Testing* **1986**, Nov., 43–44 (1986).

22. B. G. Cohen, Nondestructive evaluation of die attach bonds by IR microscopy, Technical Bulletin No. 2, Research Devices Inc., Berkeley Heights, N.J. (1986).

23. F. J. Henley *et al.*, CMOS latch-up characterization using a laser scanner, Proc. IEEE Int. Reliability Physics Symposium, 1983, pp. 122–129.

24. H. W. Marten and O. Hildebrand, Computer simulation of EBIC linescans across $p-n$ junctions, in *Scanning Electron Microscopy 1983*, pp. 1197–1209, SEM Inc., AMF O'Hare (Chicago), Ill. (1983).

25. T. Tamana and N. Kuji, Automated fault diagnostic EB tester and its application to a 40 k gate VLSI, Proc. 1985 IEEE Int. Test Conference, p. 643.

26. F. J. Henley, An automated laser prober to determine VLSI internal node logic states, Proc. 1984 IEEE Int. Test Conference, p. 536.

27. S. F. Schreiber, Beam probing for IC and wafer inspection, *Test and Measurement World* **1986**, May, 31–41 (1986).

28. D. B. Shu, C. C. Li, J. F. Mancuso, and Y. N. Sun, A line extraction method for automated SEM inspection of VLSI resist, *IEEE Trans. Pattern Anal. Mach. Intelligence* **PAMI-10**, 117–120 (1988).

29. H. Becker, High resolution metrology for submicron masks and wafers, *Microelectronic Manufacturing and Testing* **1988**, April, 11–16 (1988).

30. H. K. Nishihara and P. A. Crossley, Measuring photolithographic overlay accuracy and critical dimensions by correlating binarized Laplacians of gaussian convolutions, *IEEE Trans. Pattern Anal. Mach. Intelligence* **PAMI-10**, 17–30 (1988).

31. A. Bonora, Semiconductor wafer flatness measurements, *Test and Measurement World* **1983**, May, 57–68 (1983).

32. Certification Requirements for Hybrid Microcircuit Facilities and Lines, Govt. Printing Office, MIL-STD-1772, Washington, D.C.

33. M. T. Postek and D. C. Joy, Submicrometer microelectronics dimensional metrology, *J. Res. Nat. Bur. Stand.* **92**, 205, 1980.

34. J. M. Patternson, A noncontact voltage measurement technique using Auger spectroscopy, Proc. IEEE Int. Reliability Physics Symposium, 1983, pp. 150–153.

35. E. M. Fleuren, A very sensitive simple analysis technique using nematic liquid crystals, Proc. IEEE Int. Reliability Physics Symposium, 1983, pp. 148–149.

36. P. M. Fauchet, The Raman microprobe, *IEEE Circuits and Devices Magazine* **1986**, Jan., 37–42 (1986).

37. J. S. Blakemore *et al.* (eds.), Defect recognition and image processing in III–V compounds, in *Proc. DRIP II*, Monterey, Calif., 27–29 April 1987, Elsevier, Amsterdam (1988).

38. C. Peremarti, Microscope optique à balayage pour l'étude des composants optoélectroniques, *Appl. Opt.* **23,** 344–347 (1984).
39. G. R. Booker, Developments in semiconducting materials applications of the SEM, Institute of Physics, U.S. Govt. Printing Office 0305–2346/81/0 060–020 3, Washington D.C. (1981).
40. W. S. Andrus *et al.*, Surface analysis as a microelectronics problem solver, *Semiconductor International* **1980,** Nov., 71–84 (1980).
41. C. Pynn, *Strategies for Electronics Test*, McGraw-Hill, New York (1986), 174 pp.
42. L. F. Pau, Visual screening of ICs for metallization faults by pattern analysis methods, Proc. IEEE 1980 Int. Conf. on Cybernetics and Society, Cambridge, Mass., Oct. 1980.
43. L. F. Pau, Integrated testing and algorithms for visual inspection of semiconductor ICs, Proc. 5th Int. Joint Conf. on Pattern Recognition, Miami Beach, Fla., Dec. 1–4, 1980; IEEE Cat. 80-CH-1499-3, pp. 238–241, New York (1980).
44. L. F. Pau, Knowledge representation for sensor fusion, *Automatica* **25,** 207–214 (1989).
45. H. C. Rickers, Microcircuit screening effectiveness, Report TRS-1, Reliability Analysis Center, RADC, Griffiss AFB, N.Y. (1979).
46. D. G. Edwards, Testing for MOS IC failure modes, *IEEE Trans. Reliability* **R-31,** 9–18 (1982).
47. F. Van de Wiele (ed.), *Process and Device Modelling for IC Design* (NATO ASI Series), Noordhoff, Leyden (1977).
48. DM Date Inc., How to analyze failures of semiconductor parts, DM Data Inc. Tucson, AZ (1987).
49. M. G. Buehler, Comprehensive test patterns with modular test structures, *Solid State Technol.* **1979,** Oct., 89–94 (1979).
50. T. J. Russell and D. A. Maxwell, A production compatible microelectronic test pattern for evaluating photomask alignment, NBS Special Publ., pp. 400–451, U.S. Government Printing Office, Washington, D.C. (1979).
51. H. d'Angelo, Testing networks and the CAD of multistage screening processes, Proc. IEEE Southeast Conference, Atlanta, Ga., Apr. 10–12, 1978, pp. 213–216.
52. W. M. Vancleemput, Topological circuit layout, Report AD-A-048050, National Technical Information Service, Springfield, Va. (Oct. 1976).
53. J. D. Selim, Infrared detection of surface charge and current distribution, Report AD-A-050243, National Technical Information Service, Springfield, Va. (1979).
54. L. F. Pau and M. El Nahas, *An Introduction to Infrared Image Acquisition and Classification*, Wiley, New York (1983).
55. B. K. Horn, A problem in computer vision: Orienting silicon IC chips for lead bonding, *Comput. Graphics Image Processing* 294–303 (1975).
56. L. F. Pau, Integrated testing and algorithms for visual inspection of IC's, *IEEE Trans. Pattern Anal. Mach. Intelligence* **PAMI-5,** 602–608 (1983).
57. L. F. Pau, Method of examining and testing an electric device such as an integrated or printed circuit, U.S. Patent 4,712,057 (22/5/1984), U.K. Patent GB-0129508, FRG Patent P-34-62 182.2-08, Japan Patent Appl. 59–501908.
58. L. Beiser, Laser scanning systems, in *Laser Applications*, vol. 2, pp. 53–59, Academic Press, New York (1974).
59. E. Wolfgang *et al.*, *IEEE J. Solid State Circuits* **SC-14,** 471–481 (1979).
60. J. L. Freeman *et al.*, Direct electro-optic sampling of analog and digital GaAs ICs, Proc. 1985 GaAs IC Symposium.

61. L. F. Pau, Failure detection processes by an expert system and hybrid pattern recognition, *Pattern Recognition Lett.* **2**, 419–425 (1984).
62. Holographic wafer inspection goes on-line, *Laser Focus* **1987**, Oct., 24 (1987).
63. E. G. Meder, An algorithm to determine wafer flatness, *Semiconductor International* **1985**, July, 110–111 (1985).
64. M. Mittelstaedt, Nondestructive testing, *Photonics Spectra* **1987**, Oct., 121–130 (1987).
65. L. F. Pau, *Statistical Quality Control by Measurements for Instrumentation* (in French), Editions Chiron, Paris (1978).
66. H. Yoda, Y. Ohuchi, Y. Taniguchi, and M. Ejiri, An automatic wafer inspection system using pipelined image processing techniques, *IEEE Trans. Pattern Anal. Mach. Intelligence* **PAMI-10**, 4–16 (1988).
67. Photomask and reticle defect detection, *Semiconductor International* **1985**, April, 66–73 (1985).
68. A. C. Titus, Photomask defects: causes and solutions, *Semiconductor International* **1984**, Oct., pp. 46–52 (1984).
69. M. Shiba *et al.*, Automatic inspection of contaminants on reticles, *Proc. SPIE* **470**, 233–239 (1984).
70. G. W. Brooks *et al.*, Inspection strategies for 1 × reticles, *Semiconductor International* **1985**, April, 80–83 (1985).
71. D. G. Brobow, Expert systems, perils and promise, *Commun. ACM* **29**, 880–894 (1986).
72. D. A. Waterman, *A Guide to Expert Systems*, Addison-Wesley Publishing Co., Reading, Mass. (1986).
73. E. Rich, *Artificial Intelligence*, McGraw-Hill Book Co., New York (1986).
74. D. Petkovic and E. B. Hinkle, A rule based system for verifying engineering specifications in industrial visual inspection applications, *IEEE Trans. Pattern Anal. Mach. Intelligence* **PAMI-9**, 306–309 (1987).
75. A. Barr and E. Feigenbaum, *The Handbook of Artificial Intelligence*, W. Kaufman Publ., Los Altos, Calif. (1981).
76. A. M. Darwish and A. K. Jain, A rule based approach for visual pattern inspection, *IEEE Trans. Pattern Anal. Mach. Intelligence* **PAMI-10**, 56–68 (1988).
77. M. N. S. Swamy and K. Thulasiraman, *Graphs, Networks and Algorithms*, Wiley, New York (1981).
78. D. J. Svetkoff, J. B. Candlish, and P. W. Vanatta, High resolution imaging for the automatic inspection of multi-layer thick film circuits, Proc. 2nd Applied Machine Vision Conference, Feb. 1983.
79. C. A. Mead and L. A. Conway, *Introduction to VLSI Systems*, Addison-Wesley, Publishing Co., Reading, Mass. (1980).
80. J. Holloway, G. Steele, and G. Sussman, Scheme-79 chip, Memo 599, MIT Artificial Intelligence Lab, Cambridge, Mass. (1980).
81. Automatic inspection: the eyes don't have it, *Electronics Manufacture & Test* **1987**, May, 25–28 (1987).
82. Synervision 2000 description, VISIONETICS Corporation, Brookfield Center, Conn. (1986).
83. Y. Hara, H. Doi, K. Karasaki, and T. Iida, A system for PCB automated inspection using fluorescent light, *IEEE Trans. Pattern Anal. Mach. Intelligence* **PAMI-10**, 69–78 (1988).
84. J. M. Kallis, L. A. Strattan, and T. T. Bui, Programs help spot hot spots, *IEEE Spectrum* **1987**, March, 36–41 (1987).

85. W. Perry et al., U.S. Patent 4,165,170, 21 Aug. 1979.

86. C. L. Sørensen, Quality control of multilayer PCB, in *Reliability in Electronics and Electronic Components and Systems* (E. Lauger, ed.), pp. 887–889, North-Holland, Amsterdam (1982).

87. S. Kahan, T. Pavlidis, and H. S. Baird, On the recognition of printed characters of any font and size, *IEEE Trans. Pattern Anal. Mach. Intelligence* **PAMI-9**, 274–288 (1987).

88. T. Adams, Nondestructive inspection of hybrid circuits using the scanning laser acoustic microscope, *Hybrid Circuit Technology* **1984**, July, pp. 71–73 (1984).

89. M. Berger, *Neutron Radiography*, Elsevier, Amsterdam (1965).

90. F. N. Bradley, Ultrasonic scanning of multilayer ceramic chip capacitors, AVX Corp., Myrtle Beach, S.C. (1982).

91. G. R. Love, Nondestructive testing of monolithic ceramic capacitors, Proc. 1973 ISHM Conference.

92. Linewidth measurement aids process control, *Semiconductor International* **1985**, Feb., 66–73 (1985).

93. M. Rioux, Laser range finder based on synchronized scanners, *Appl. Opt.* **23**, 3837 (1984).

94. W. Snyder and G. Bilbro, Segmentation of 3-D images, Proc. IEEE Conf. on Robotics and Automation, 1985, p. 396.

95. S. L. Bartlett, P. J. Besl, C. L. Cole, R. Jain, D. Mukherjee, and M. D. Skifstad, Automatic solder joint inspection, *IEEE Trans. Pattern Anal. Mach. Intelligence* **PAMI-10**, 31–43 (1988).

96. R. Vanzetti, A. C. Traub, and J. S. Ele, Hidden solder joint defects detected by laser infrared system, Proc. IPC 24th Ann. Meeting, 1981, pp. 1–15.

97. M. P. Seah and C. Lea, Certainty of measurement using an automated infrared laser inspection instrument for PCB solder joint integrity, *J. Phys. E: Sci. Instrum.* **18**, 676–682 (1985).

98. P. J. Klass, Texas Instruments tests laser to detect and repair faulty soldering, *Aviation Week and Space Technology* **1985**, Sept. 30, 98–100 (1985).

99. D. W. Capson and S. K. Eng, A color illumination approach for machine inspection of solder joints, *IEEE Trans., Pattern Anal. Mach. Intelligence* **PAMI-10**, 387–393 (1988).

100. T. H. Ooi et al., An expert system for production scheduling of IC's, Nanyang Technological Institute, Singapore (1988).

101. Pattern examining method for printed circuit manufacture, British Patent GB-2139-754-A, Japan Patent JP-0809 54, 11.05.1983.

102. P. Mansbach, Calibration of a camera and light source by fitting to a physical model, *Computer Vision, Graphics and Image Processing* **35**, 200–219 (1986).

103. K. S. Fu, *Syntactic Pattern Recognition*, Academic Press, New York (1980).

104. A. Rozenfeld and A. K. C. Kak, *Digital Image Processing*, 2nd Ed., Academic Press, New York (1983).

105. A. K. Jain, *Fundamentals of Digital Image Processing*, Prentice-Hall, Inc., Englewood Cliffs, N.J. (1988).

106. F. A. Jenkins and H. E. White, *Fundamentals of Optics*, McGraw-Hill Book Co., New York (1976).

107. K. Fukunaga, *Introduction to Statistical Pattern Recognition*, Academic Press, New York (1972).

108. S. Even, *Graph Algorithms*, Computer Science Press, Rockville, Md. (1979).

109. E. C. Driscoll, Fast image comparison using a simplified rank correlation algorithm, *Proc. SPIE* **757**, Paper 23 (1986).

110. B. G. Batchelor, D. A. Hill, and D. C. Hodgson, *Automated Visual Inspection*, North-Holland Publishing Co./IFS Publications, Amsterdam (1985).

111. D. J. Hall, R. M. Endlich, D. E. Wolf, and A. E. Brain, Objective methods for registering landmarks and determining cloud motions from satellite data, *IEEE Trans. Comput.* **C-21**, 768–776 (1972).

112. D. I. Barnea and H. F. Silverman, A class of algorithms for fast digital image registration, *IEEE Trans. Comput.* **C-21**, 179–186 (1972).

113. M. Onoe and M. Saito, Automatic threshold setting for the sequential similarity detection algorithm, *IEEE Trans. Comput.* **C-25**, 1052–1053 (1976).

114. P. W. Woods, C. J. Taylor, and R. Wicht, The use of geometric and grey-level models for industrial inspection, *Pattern Recognition Lett.* **5**, 11–17 (1987).

115. L. S. Davis, A survey of edge detection technique, *Comput. Graphics Image Processing* **4**, 248–270 (1975).

116. L. G. Roberts, Machine perception of three dimensional solids, in *Optical Electro-optical Processing of Information*, pp. 159–197, MIT Press, Cambridge (1965).

117. J. M. S. Prewitt, Object enhancement and extraction, in *Picture Processing and Psychopictorics*, pp. 75–149, Academic Press, New York (1970).

118. R. O. Duda and P. E. Hart, *Pattern Classification and Scene Analysis*, Wiley, New York (1971), pp. 267–272.

119. R. Kirsch, Computer determination of the constituent structure of biological images, *Comput. Biomed. Res.* **4**, 315–328 (1971).

120. G. S. Robinson, Edge detection by compass gradient masks, *Comput. Graphics Image Processing* **6**, 492–501 (1977).

121. M. H. Hückel, An operator which locates edges in digitized pictures, *J. ACM* **18**, 113–125 (1971).

122. M. H. Hückel, A local visual operator which recognizes edges and lines, *J. ACM* **20**, 634–647 (1973); erratum in *J. ACM* **21**, 350 (1974).

123. L. Mero and Z. Vassy, A simplified and fast version of the Hueckel operator for finding optimal edges in pictures, Proc. 4th Int. Joint Conf. on Artificial Intelligence, 1978, pp. 650–655.

124. F. O'Gorman, Edge detection using Walsh function, *Artificial Intelligence* **10**, 215–223 (1978).

125. J. W. Modestino and R. W. Fries, Edge detection in noisy images using recursive filtering, *Comput. Graphics Image Processing* **6**, 409–433 (1977).

126. T. Kasvand, Iterative edge detection, *Comput. Graphics Image Processing* **4**, 279–286 (1975).

127. R. O. Duda and P. E. Hart, Use of the Hough transformation to detect lines and curves in pictures, *Commun. ACM* **15**, 11–15 (1972).

128. F. O'Gorman and M. Clowes, Finding picture edges through collinearity of feature points, *IEEE Trans. Comput.* **C-25**, 449–456 (1976).

129. A. Martielli, An application of heuristic search methods to edge and contour detection, *Commun. ACM* **19**, 73–83 (1976).

130. A. V. Oppenheim and R. W. Schafer, *Digital Signal Processing*, Prentice-Hall, Englewood Cliffs, N.J. (1975).

131. S. N. Lapidus, Understanding how images are digitized, in *Proceedings of the 2nd Conference on Machine Intelligence*, IFS Publications, Bedford, U.K. (1985).

132. J. N. Gupta and P. A. Wintz, A boundary finding algorithm and its application, *IEEE Trans. Circuits Syst.* **CAS-22**, 351–362 (1975).

133. C. R. Brice and C. L. Fennema, Scene analysis using regions, *Artificial Intelligence* **1**, 205–226 (1970).

134. T. Pavlidis, Segmentation of pictures and maps through functional approximation, *Comput. Graphics Image Processing* **1**, 360–372 (1972).

135. S. L. Horowitz and T. Pavlidis, Picture segmentation by a directed split-and-merge procedure, Proc. 2nd Int. Joint Conf. on Pattern Recognition, 1974, pp. 424–433.

136. P. M. Narendra and M. Goldberg, A graph-theoretic approach to image segmentation, Proc. IEEE Conf. on Pattern Recognition and Image Processing, 1977, pp. 248–256.

137. T. Asano and N. Yokoya, Image segmentation scheme for low-level computer vision, *Pattern Recognition* **14**, 267–273 (1981).

138. N. Yokoya, T. Kitahashi, K. Tanaka, and T. Asano, Image segmentation scheme based on a concept of relative similarity, Proc. 4th Int. Joint Conf. on Pattern Recognition, 1978, pp. 645–647.

139. Y. Noguchi, Y. Tenjin, and T. Sugishita, A segmentation method of cellular images using color information (1), *Bull. Electrotech. Lab.* **41**, 38–63 (1977).

140. G. B. Coleman and H. C. Andrews, Image segmentation by clustering, *Proc. IEEE* **67**, 773–785 (1979).

141. N. Otsu, A threshold selection method from gray-level histograms, *IEEE Trans. Syst. Man, Cybern.* **SMC-9**, 62–66 (1979).

142. G. Nagy and J. Tolaba, Nonsupervised crop classification through airborne multi-spectral observation, *IBM J. Res. Develop.* **16**, 138–153 (1972).

143. R. M. Haralick and I. Dinstein, A spatial clustering procedure for multi-image data, *IEEE Trans. Circuits Syst.* **CAS-22**, 440–449 (1975).

144. Y. Fukada, Spatial clustering procedures for region analysis, Proc. 4th Int. Joint Conf. on Pattern Recognition, 1978, pp. 329–331.

145. J. A. Feldman and Y. Yakimovsky, Decision theory and artificial intelligence: I. A semantics-based region analyzer, *Artificial Intelligence* **5**, 349–371 (1974).

146. J. M. Tenenbaum and H. G. Barrow, IGS: A paradigm for integrating image segmentation and interpretation, Proc. 3rd Int. Joint Conf. on Pattern Recognition, 1976, pp. 504–513.

147. J. Serra, *Image Analysis and Mathematical Morphology*, Academic Press, London (1982).

148. R. Thibadeau, Printed circuit board inspection, CMU-RI-TR-81-8, The Robotics Institute, Carnegie-Mellon University, Pittsburgh (1981), 29 pp.

149. T. N. Mudge, R. A. Rutenbar, R. M. Lougheed, and D. E. Atkins, Cellular image processing techniques for VLSI circuit layout validation and routing, Proc. ACM 19th Design Automation Conf., June 1982.

150. A. Rosenfeld and J. L. Pfaltz, Sequential operations in digital picture processing, *J. ACM* **13**, 471–494 (1966).

151. A. Toriwaki and S. Yokoi, Distance transformations and skeletons of digitized pictures with applications, in *Progress in Pattern Recognition* (L. Kanal and G. Genselma, eds.), Vol. 1, pp. 187–264, North-Holland, Amsterdam (1981).

152. S. Yokoi, J. Toriwaki, and T. Fukumura, A sequential algorithm for shrinking binary pictures, *J. Syst., Comput., Controls* **10**, 69–95 (1979).

153. S. Levialdi, On shrinking binary picture patterns, *Commun. ACM*, **15**, 7–10 (1972).

154. C. V. K. Rao, B. Prasada, and K. R. Sarma, A parallel shrinking algorithm for binary patterns, *Comput. Graphics Image Processing* **5**, 265–270 (1976).

155. B. Carré, *Graphs and Networks*, Clarendon, Oxford (1979).

156. R. E. Tarjan, A note on finding the bridges of a graph, *Inform. Processing Lett.* **2,** 160–161 (1974).

157. M. K. Hu, Invariant image moments, *IEEE Trans. Inform. Theory* **IT-8,** 179 (1962).

158. G. B. Curevich, *Foundations of the Theory of Algebraic Invariants,* P. Noordhoff, Groningen, Holland (1964).

159. S. Dudani, *IEEE Trans. Comput.* **C-26,** 39 (1977).

160. D. Casadent and D. Psaltis, Hybrid processor to compute invariant moments for pattern recognition, *Opt. Lett.* **5,** 395–397 (1980).

161. C. T. Zahn and R. Z. Roskies, Fourier descriptors for plane closed curves, *IEEE Trans. Comput.* **C-21,** 269–281 (1972).

162. D. Dubois and H. Prade, *Fuzzy Sets and Systems,* Academic Press, New York (1981).

163. E. T. Lee and L. A. Zadeh, Note on fuzzy languages, *Inform. Sci.* **1,** 421–434 (1969).

164. R. M. Haralick, K. Shanmugam, and I. Dinstein, Texture features for image classification, IEEE Trans. *Syst., Man, Cybern.* **SMC-3,** 610–621 (1973).

165. H. Freeman, Computer processing of line-drawing images, *ACM Comput. Survey,* **6,** 57–97 (1974).

166. AVA Machine Vision Glossary, Society of Manufacturing Engineers, Dearborn, MI (1985).

167. D. O'Shea, *Elements of Modern Optical Design,* Wiley, New York (1985).

Suggested Readings

168. W. Frei and C. C. Chen, Fast boundary detection: A generalization and a new algorithm, *IEEE Trans. Comput.* **C-26,** 988–998 (1977).
169. Y. Ohta, T. Kanade, and T. Sakai, Color information for region segmentation, *Comput. Graphics Image Processing* **13,** 222–241 (1980).
170. M. Yachida and S. Tsuji, A versatile machine vision system for complex industrial parts, *IEEE Trans. Comput.* **C-26,** 882–894 (1977).
171. M. Nagase, *Microelectronics and Reliability* **20,** 717–735 (1980).
172. L. F. Pau, Approaches to industrial image processing and their limitations, *IEE J. Electronics Power* **1984,** Feb, 135–140 (1984).
173. C. H. Lane, Considerations in microcircuit visual inspection, Report RADC-TR-75-150, National Technical Information Service, Springfield, Va. (June 1975).
174. R. P. Kruger and W. B. Thomson, A technical and economic assessment of computer vision for industrial inspection and robotic assembly, *Proc. IEEE* **69,** 1524–1538 (1981).
175. M. J. Howes and D. V. Morgan, *Reliability and Degradation in Semiconductor Devices and Circuits,* Wiley, London (1981).
176. J. F. Jarvis, A method for automating the visual inspection of printed wiring boards, *IEEE Trans. Pattern Anal. Mach. Intelligence* **PAMI-2,** 77–82 (1980).
177. J. F. Jarvis, Visual inspection automation, *IEEE Computer* May, **13,** 32–38 (1980).
178. K. S. Fu (ed.), Special issue on robotics and automation, *IEEE Computer* **15,** Dec. (1982).
179. R. T. Chin and C. A. Harlow, Automated visual inspection: A survey, *IEEE Trans. Pattern Anal. Mach. Intelligence* **PAMI-4,** 557–573 (1982).
180. J. T. Healy, *Automatic Testing and Evaluation of Digital Integrated Circuits,* Prentice-Hall, Englewood Cliffs, N.J. (1980).
181. F. Jensen and N. E. Petersen, *Burn-In,* Wiley, London (1982).
182. L. F. Pau, *Failure Diagnosis and Performance Monitoring,* Marcel Dekker, New York (1981).
183. ICE Corporation, Visual inspection criteria for ICs and hybrids, MIL-STD-883-C, Method 2010/2017, ICE Corporation, Scottsdale, Ariz. (Oct. 1987).
184. Properties of silicon, *EMIS Datareviews,* No. 4, INSPEC, IEE, London (1988).

185. Properties of gallium arsenide, *EMIS Datareviews*, No. 2, INSPEC, IEE, London (1986).
186. Properties of mercury cadmium telluride, *EMIS Datareviews*, No. 3, INSPEC, IEE, London (1987).
187. C. L. Sørensen, Quality control of multilayer PCBs and electric components, in *Reliability in Electrical and Electronic Components and Systems* (E. Lauger, ed.), North-Holland Amsterdam (1982), pp. 474–476.
188. S. F. Schreiber, Beam probing for IC and wafer inspection, *Test and Measurement World* **1986,** May, 32 (1986).
189. J. M. Kallis and T. T. Bui, Programs help spot hot spots, *IEEE Spectrum* **1987,** March, **24,** 36–41 (1987).
190. W. M. Sterling, Nonreference inspection of complex and repetitive patterns, *Proc. SPIE*, **282,** 182–190 (1981).
191. R. C. Restrik, An automatic optical PCB inspection system, *Proc. SPIE*, **116,** 76–81 (1977).
192. J. P. Donahue and C. W. Souder, 3-D laser scanners for SMT and hybrid circuit inspection, SME Paper MS 86-613, SME, Dearborn, Mich. (1986).
193. R. Cernera and R. Ingalls, Evaluating optical linewidth measurement systems, *Microelectronic Manufacturing and Testing* **9,** July (1986).
194. S. J. Erasmus, Inspection techniques for semiconductor lithography, Hewlett-Packard Co., Palo Alto, Calif. (1982).
195. P. Burggraaf, Wafer inspection for defects, *Semiconductor International* **1985,** July, 55–65 (1985).
196. D. H. Ballard and C. M. Brown, *Computer Vision*, Prentice-Hall, Englewood Cliffs, N.J. (1986).
197. B. K. P. Horn, *Robot Vision*, MIT Press, Cambridge, Mass. (1986).
198. B. di Bartolo (ed.), *Spectroscopy of Solid State Laser Type Materials,* Plenum, New York (1988).
199. W. Hastie, Bare board testing, *Circuits Manufacturing* **22**(1), 36–48 (1982).
200. G. Koutures, Automated optical inspection paces future PC production, *Electronic Packaging & Production* **23**(6), 120–128 (1983).
201. C. Coombs (ed.), *Printed Circuits Handbook,* 2nd Ed., McGraw-Hill Book Co., New York (1979), 634 pp.
202. Printed circuits fundamentals, *Electronics,* **29**(2), 25–63 (1983).
203. R. Bupp, L. Chellis, R. Ruane, and J. Wiley, High-density board fabrication techniques, *IBM J. Res. Develop.* **26,** 297–305 (1982).
204. Specification for printed circuits of assessed quality: Generic data and methods of test: Capability approval procedure and rules, BS 9760:1977, British Standard Institution, London (1977), 45 pp.
205. Guidelines for acceptability of printed boards, ANSI/IPC-A-600-C-1978, The Institute for Interconnecting and Packaging Electronic Circuits, Evanston, Ill. (1978), 79 pp.
206. Specification for printed circuits of assessed quality: Sectional specification for double-sided printed circuits with plated through holes, BS 9762:1977, British Standard Institution, London (1977), 50 pp.
207. Specification for printed circuits of assessed quality: Sectional specification for multilayer printed circuits, BS 9761:1977, British Standard Institution, London (1977), 49 pp.
208. M. West, S. DeFoster, E. Baldwin, and R. Ziegler, Computer-controlled optical testing of high density printed circuit boards, *IBM J. Res. Develop.* **27,** 50–58 (1983).

209. B. W. Hueners and D. M. Day, Advances in automated wire and die bonding, *Solid State Technol.* **1983,** March, **14,** 69–76 (1983).

210. The Optrotech Automatic P.C.B. Flaw Detection & Designation System, Optrotech Ltd., Nes Ziona, Israel (1983), 6 pp.

211. W. Bentley, The Inspectron: An automatic optical printed circuit board (PCB) inspector, Proc. SPIE Conf. on Optical Pattern Recognition, Bellingham, 1979, pp. 37–109.

212. W. Sterling, Automatic non-reference inspection of printed wiring boards, Proc. IEEE Conf. on Pattern Recognition and Image Processing (PRIP), Chicago, 6–8 August 1979, pp. 93–100, IEEE, New York (1979).

213. Integrated circuit metrology, inspection and process control, *Proc. SPIE* **775** (1987).

214. N. Goto and T. Kondo, An automatic inspection system for printed wiring board masks, *Pattern Recognition,* **12,** 443–455 (1980).

215. Y. Hara, K. Okumoto, T. Hamada, and N. Akiyama, Automatic visual inspection of LSI photomasks, Proc. 5th Int. Conf. on Pattern Recognition, Dec. 1980, pp. 273–279.

216. Automatic reticle/mask inspection system for VLSI, *Solid State Technol.* **26**(1), 45–46 (1983).

217. Chipcheck reticle/mask inspection system, *Solid State Technol.* **26**(5), 66 (1983).

218. B. Tsujiyama, K. Saito, and K. Kurihara, A highly reliable mask inspection system, *IEEE Trans. Electron Devices* **ED-27,** 1284–1290 (1980).

219. Infrared thermal measurements on microelectronic circuits, *Solid State Technol.* April, **17,** 79–81 (1986).

220. P.-E. Danielsson, Encoding of binary images by raster-chain-coding of cracks, Proc. 6th Int. Joint Conf. on Pattern Recognition, Munich, 1982, pp. 335–338.

221. J.-C. Pineda and P. Horaud, An improved method for high-curvature detection with applications to automated inspection, *Signal Processing* **5**(2), 117–125 (1983).

222. Printed board description in digital form, ANSI/IPC-D-350B, The Institute for Interconnecting and Packaging Electronic Circuits, Evanston, Ill. (1977). 29 pp.

223. L. Uhr (ed.), *Parallel Computer Vision,* Academic Press, New York (1987).

224. R. Jarvis, A perspective on range finding techniques for computer vision, *IEEE Trans. Pattern Anal. Mach. Intelligence* **PAMI-5,** 122 (1983).

225. U. Montanari, Continuous skeletons from digitized images, *J. ACM* **16,** 534–549 (1969).

226. H. Hinkelmann, Inducing photocurrents in IC chips with the laser scan microscope, Proc. 5th Annual Test and Measurement World Expo., 1986, p. 219.

227. J. W. Ruch, Pattern verification through die to data base inspection, *Proc. SPIE* **772,** Paper 772–32 (1987).

228. Integrated Circuit Metrology, Inspection and Process Control, *Proc. SPIE* **775** (1987).

229. A. Thaer, Akustomikroskop ELSAM, *Leitz Mitt. Wiss., Techn.* **8**(3/4), May, 61–67 (1982).

230. C. F. Quate, Microwaves, acoustics and scanning microscopy in *Scanned Image Microscopy* (E. A. Ash, ed.), pp. 23–55, Academic Press, London, (1980).

231. B. W. Hueners and D. M. Day, Advances in automated wire and die bonding, *Solid State Technol.* **1983,** March, 69–76 (1983). (See Ref. 209).

232. D. M. Domres and J. MacFarlane, Automatic optical inspection techniques for PWBs: image acquisition and analysis remain areas of challenge, *Test and Measurement World* **1983,** May, 71–77 (1983).

233. S. Corman, Optical inspection technology for multi-layer PWB, *Test and Measurement World* **1983,** May, 78–86 (1983).

234. Program to automate printed circuit board inspection, Report ECC-81-132 R-2, SRI International, Menlo Park, CA (1982).

235. M. Ejiri and T. Uno, A process for detecting defects in complicated patterns, *Comput. Graphics Image Processing* **2**, 326–339 (1973).

236. Ultrafast Laser Probe Phenomena in Bulk and Microstructure Semiconductors II, *Proc. SPIE* **942** (1988).

237. E. A. Giess, Magnetic bubble materials, *Science* **208**, 938–943 (1980).

238. F. P. Andresen, S. D. Roth, and B. E. Shimano, Rulers: A simple and versatile inspection tool, SME, Dearborn, MI (1987).

239. V. C. Bolhouse, Machine vision automates inspection of thick film hybrids, *IEEE Circuits and Devices Magazine* **1986**, Jan., 44–48 (1986).

240. S. H. Mersch, Polarized lighting for machine vision applications, SME Proceedings, 3rd Applied Machine Vision Conf., Feb. 1984, pp. 4–40.

241. A. Fridge, J. Lee, and R. Walker, Automated optical inspection of hybrid circuitry (published reference unknown).

242. E. Menzel and R. Buchanan, Electron beam probing of integrated circuits, *Solid State Technol.* **1985**, Dec. (1985).

243. F. L. Skaggs, Machine vision system speeds PCB inspection and increases measurement efficiency, *Test and Measurement World* **1984**, June, 94–104 (1984).

244. *Semiconductor International* **1985**, April, 80 (1985).

245. Special issue on industrial machine vision and computer technology, Part I, *IEEE Trans. Pattern Anal. Mach. Intelligence* **PAMI-10**(1), (1988) Part II, **PAMI-10**(3), (1988).

246. J. K. Chou, Automated inspection of printed circuit board, SME Vision '85 Conf. Proc., March 1985, pp. 5.20–5.33.

247. N. Goto and T. Kondo, An automatic inspection system for printed wiring board masks, *Pattern Recognition* **12**, 443–455 (1980).

248. Y. Hara, N. Akiyama, and K. Karasaki, Automatic inspection system for printed circuit boards, *IEEE Trans. Pattern Anal. Mach. Intelligence* **PAMI-5**, 623–630 (1983).

249. P. Kaufmann, G. Medioni, and R. Nevatia, Visual inspection using linear features, *Pattern Recognition,* **17**, 485–491 (1984).

250. G. A. W. West, A system for the automatic visual inspection of bare-printed circuit boards, *IEEE Trans. Syst. Man, Cybern.* **SMC-14**, 767–773 (1984).

251. M. A. West, S. M. DeFoster, E. C. Baldwin, and R. A. Ziegler, Computer-controlled optical testing of high-density printed-circuit boards, *IBM J. Res. Develop.* **27**, 50–58 (1983).

252. A. J. Billota, *Connections in Electronic Assemblies,* Marcel Dekker, New York (1985), pp. 1–72.

253. C. E. Jowett, *Reliable Electronic Assembly Production,* Business Books Ltd., London (1970), pp. 53–84.

254. H. H. Manko, *Soldering Handbook for Printed Circuits and Surface Mounting,* Van Nostrand Reinhold, New York (1986).

255. Y. Nakagawa, Automatic visual inspection of solder joints on printed circuit boards, *Robot Vision* **336**, 121–127 (1982).

256. W. E. McIntosh, Automating the inspection of printed circuit boards, *Robotics Today* **1983**, June, 75–78 (1983).

257. D. W. Capson, An improved algorithm for the sequential extraction of boundaries from a raster scan, *Computer Vision, Graphics, and Image Processing* **28**, 109–125 (1984).

258. A. M. Wallace, Industrial applications of computer vision since 1982, *IEE Proceedings* **135,** Part E, No. 3, 117–136 (1988) (with 263 references).

259. Scanning electron microscopy for production, *Semiconductor International* **1987,** Aug., 51–58 (1987).

260. D. Jacobus, X-ray template, *PC Fab.* **1987,** March, 36–39 (1987).

261. ICE Corporation, Practical IC Fabrication, ICE Corporation, Scottsdale, Ariz. (1987).

262. B. Davis, *The Economics of Automatic Testing,* McGraw-Hill, New York (1982).

263. Very high magnification optical laser microscope, *Solid State Technol.* **1986,** April, 67–69 (1986).

264. J. Bretschi, *Automated Inspection Systems for Industry,* IFS Publications Ltd., Bedford, U.K. (1981).

265. F. L. Skaggs, Machine vision system speeds PCB inspection and increases measurement efficiency, *Test and Measurement World* **1984,** June, 94–111 (1984).

266. I. Morishita and M. Okumura, Automated Visual Inspection Systems for Industrial Applications, in *Proceedings of the 9th IMEKO Congress of the International Measurement Confederation, Berlin,* Vol. 3, North-Holland, Amsterdam (1983).

267. R. L. Seliger, Ion beams promise practical systems for submicrometer wafer lithography, *Electronics* **1980,** March 27 (1980), p. 48.

268. E. Menzel and R. Buchanan, Electron beam probing of integrated circuits, *Solid State Technol.* **1985,** Dec., 63–70 (1985).

269. E. Doyle and W. Morris, *RADC Microelectronics Failure Analysis Procedural Guide,* Reliability Analysis Center, Rome, N.Y., Catalogue MFAT-1.

270. Nondestructive tests used to insure the integrity of semiconductor devices, with emphasis on acoustic emission techniques, NBS Special Publication 400-59, National Bureau of Standards, Washington, D.C. (1981).

271. C. G. Masi, Electronic inspection in the early 1990's, *Test and Measurement World* **1987,** Dec., 68–78 (1987).

272. W. P. Brennan, Applications of thermal analysis in the electrical and electronics industries (thermal properties of various substrates), Thermal Analysis Application Study No. 25, Perkin-Elmer Instrument Div., Norwalk, Conn. (1978).

273. Pattern recognition on bonders and probers, *Semiconductor International* **1981,** Feb., 53–70 (1981).

274. S. L. Bartlett, C. L. Cole, and R. Jain, Expert system for visual solder joint inspection, IEEE Conf. on Artificial Intelligence Applications, March 1987.

275. P. G. Besl and R. Jain, Invariant surface characteristics for 3-D object recognition in depth maps, *Computer Vision, Graphics and Image Processing* **33,** 33–80 (1986).

276. P. B. Besl, E. Delp, and R. Jain, Automatic visual solder joint inspection, *IEEE J. Robotics and Automation* **1,** 42–56 (1985).

277. M. Brenner, Computer enhancement of low-light microscopic images, *Am. Clin. Prod. Rev.* **32,** No. 2(5) (1983).

278. O. Silven *et al.,* A design data based visual inspection system for printed wiring, in *Applications of Image Processing* (J. Sanz, ed.), Springer-Verlag, Heidelberg (1989), Chap. 5, pp. 112–126.

279. S. Selberherr, *Analysis and Simulation of Semiconductor Devices,* Springer-Verlag, Vienna (1984).

280. S. C. Seth and V. D. Agrawal, Cutting chip testing costs, *IEEE Spectrum* **1985,** April, 38–45 (1985).

281. C. H. Mangin and S. McClell, *Surface Mount Technology,* CEERIS International, IFS Publications, Bedford, U.K. (1987).

282. I. D. Ward, Analysis methods complement each other in surface studies, *Industrial R & D 1983,* Sept. (1983), pp. 121–124.

283. J. L. C. Sanz and A. K. Jain, Machine vision techniques for inspection of printed wiring boards and thick film circuits, *J. Opt. Soc. Am.* **3,** 1465–1483 (1986).

284. Standard practice for controlling quality of radiographic testing of electronic devices, No. E 801-81-A, American Society for Testing and Materials, Philadelphia, Pa. (1974).

285. M. Born and E. Wolf, *Principles of Optics,* Pergamon, Oxford (1965).

286. G. A. Boutry, *Instrumental Optics,* Interscience, New York (1965).

287. E. B. Brown, *Modern Optics,* Van Nostrand Reinhold, New York (1965).

288. R. H. Clark, *Handbook of Printed Circuit Manufacturing,* Van Nostrand Reinhold, New York (1985).

289. Very high magnification optical laser microscope, *Solid State Technol.* **1986,** April, 67–68 (1986).

290. J. N. Zemel (ed.), *Non-destructive Evaluation of Semiconductor Materials and Devices (NATO ASI Series,* No. B-46), Plenum, New York (1979).

Index